Preface

Global optimization is concerned with the characterization and computation of global minima or maxima of nonlinear functions. The general constrained global minimization problem has the following form:

Given: $K \subseteq R^n$ *compact set,* $f: K \rightarrow R$ *continuous function*

Find: $x^* \in K$, $f^* = f(x^*)$ *such that* $f^* \leq f(x)$ *for all* $x \in K$

Such problems are widespread in mathematical modeling of real world systems for a very broad range of applications. Such applications include economies of scale, fixed charges, allocation and location problems, quadratic assignment and a number of other combinatorial optimization problems. More recently it has been shown that certain aspects of VLSI chip design and database problems can be formulated as constrained global optimization problems with a quadratic objective function. Although standard nonlinear programming algorithms will usually obtain a local minimum to the problem, such a local minimum will only be global when certain conditions are satisfied (such as f and K being convex). In general several local minima may exist and the corresponding function values may differ substantially. The problem of designing algorithms that obtain global solutions is very difficult, since in general, there is no local criterion for deciding whether a local solution is global.

Active research during the past two decades has produced a variety of methods for finding constrained global solutions to nonlinear optimization problems. In this monograph we consider deterministic methods which include those concerned with enumerative techniques, cutting planes, branch and bound, bilinear programming, general approximate algorithms and large-scale approaches.

There has been a significant recent increase in research activity on the subject of constrained global optimization and related computational algorithms. This monograph summarizes much of this recent work and contains an extensive list of references to papers on constrained global optimization, deterministic solution methods, and applications. We hope that this work will be valuable for other researchers in global optimization.

We wish to express our appreciation and thanks to Andrew T. Phillips and Nainan Kovoor who carefully read an earlier version of this monograph and suggested a number of valuable improvements. The authors' research described in this monograph was supported in part by the National Science Foundation Grant DCR8405489.

June 1987 P.M. Pardalos, J.B. Rosen

Lecture Notes in Computer Science

Edited by G. Goos and J. H

268

P. M. Pardalos
J. B. Rosen

Constr
Global
Algorith

Springer-V

Berlin Heidelbe

Authors

Panos M. Pardalos
Computer Science Department, The Pennsylvania State University
University Park, PA 16802, USA

J. Ben Rosen
Computer Science Department, University of Minnesota
Minneapolis, MN 55455, USA

CR Subject Classification (1987): G.1.6, G.1.2, F.2.1

ISBN 3-540-18095-8 Springer-Verlag Berlin Heidelberg New York
ISBN 0-387-18095-8 Springer-Verlag New York Berlin Heidelberg

Printed in Germany

Printing and binding: Druckhaus Beltz, Hemsbach/Bergstr.
2145/3140-543210

Table of Contents

Chapter 1 Convex sets and functions

Convex sets and functions play a dominant role in optimization and some of their properties are essential in the study of several algorithms. In this introductory chapter we start with summary of some of the most important properties that we are going to use.

1.1 Convex sets

A subset C of the Euclidean space R^n is said to be convex if for every $x_1, x_2 \in C$ and λ real, $0 \leq \lambda \leq 1$, the point $\lambda x_1 + (1-\lambda)x_2 \in C$.

The geometric interpretation of this definition is clear. For any two points of C, the line segment joining these two points lies entirely in C.

Given the vectors $x_1, \ldots, x_m$ in R^n and real numbers $\lambda_i \geq 0$ with $\sum_{i=1}^{m} \lambda_i = 1$, the vector sum $\lambda_1 x_1 + \cdots + \lambda_m x_m$ is called a convex combination of these points. Some properties of convex sets are summarized in the next theorems.

Theorem 1.1.1: A subset of R^n is convex iff it contains all the convex combinations of its elements.

Proof: Let C be a convex set in R^n. If $x_i \in C$ and $\lambda_i \geq 0$, $i=1, \ldots, k$, such that $\sum_{i=1}^{k} \lambda_i = 1$, prove by induction on k that the convex combination $\sum_{i=1}^{k} \lambda_i x_i \in C$.

Theorem 1.1.2: Let F be a family of convex sets. Then the intersection $\bigcap_{C \in F} C$ is also a convex set.

However, it is easy to see that the union of convex sets need not be convex. Some other algebraic set operations that preserve convexity are defined below.

Theorem 1.1.3: 1) Let C be a convex set in R^n and a a real number. Then the set $aC = \{x: x = ay, y \in C\}$ is also convex.
2) Let C_1, C_2 be convex sets in R^n. Then the set

$$C_1 + C_2 = \{x: x = x_1 + x_2, x_1 \in C_1, x_2 \in C_2\}$$

is also convex.

1.2 Linear and affine spaces

For any $x, y \in R^n$ the inner product $x^T y$ is the real number $\sum_{i=1}^{n} x_i y_i$. The Euclidean norm is defined to be $\|x\| = (\sum_{i=1}^{n} x_i^2)^{1/2}$. Other notations and terminology not defined here are the standard ones used in the literature.

A hyperplane H in R^n is a set of the form

$$H = \{x \in R^n: c^T x = b\}.$$

for some vector $c \in R^n$ and $b \in R$. Similarly we define the closed half spaces

$$H_1 = \{x \in R^n: c^T x \geq b\}, \quad H_2 = \{x \in R^n: c^T x \leq b\}.$$

It is very easy to see that H, H_1, H_2 are all convex sets.

A nonempty subset V of R^n is called a (linear) subspace if the following conditions are true:
i) if $x, y \in V$ then $x+y \in V$,
ii) if $x \in V$ and $r \in R$ then $rx \in V$.

Next we define the structure of linear subspaces. Let $S = \{v_1, \ldots, v_m\}$ be a set of vectors in V. We say that S spans V if for every vector $v \in V$, $v = \sum_{i=1}^{m} c_i v_i$, where the c_i's are real numbers. The set S is said to be linearly independent if we cannot find constants $c_1, \ldots, c_m$ not all zero such that $c_1 v_1 + \cdots + c_m v_m = 0$ (otherwise S is called linearly dependent).

If the set S spans V and is linearly independent we call it a basis of V. The dimension of the subspace V, $dim(V)$, is defined to be the number of vectors in some basis S.

To get more insight into the geometric and algebraic nature of a subspace we equivalently define a linear subspace to be the set

$$V = \{x \in R^n : c_{i1}x_1 + \cdots + c_{in}x_n = 0,\ i = 1, \ldots, m\},$$

that is, V is the solution set of the homogeneous system of linear equations $Cx = 0$ where C is the $m \times n$ matrix of the coefficients c_{ij}. Here the dimension of V is equal to $n - rank(C)$ where the *rank* of the matrix is the maximum number of linearly independent columns (or rows) of the matrix.

An affine subspace A of R^n is a linear subspace V translated by some vector u, that is $A = \{x \in R^n : x = u + v,\ v \in V\}$. Also $dim(A) = dim(V)$. Equivalently we can define

$$A = \{x \in R^n : c_{i1}x_1 + \cdots + c_{in}x_n = b,\ i = 1, \ldots, m\},$$

that is, A is the solution set of the (nonhomogeneous) linear system $Cx = b$.

From the above discussion it is clear that a hyperplane in R^n is an affine subspace of dimension $n-1$.

1.3 Convex hull

Another important concept in convexity is that of forming the smallest convex set containing a given subset S in R^n. The convex hull of S is the set

$$Co(S) = \cap\{C : C \text{ convex in } R^n \text{ and } C \supseteq S\}.$$

The convex hull of a finite set of points is called a convex polytope.

It is clear that a convex polytope is always bounded. A convex polytope that contains all its boundary points is closed. A point x on the boundary of S is called an extreme point (or vertex) if there are no distinct points $x_1, x_2 \in S$ such that $x = \lambda x_1 + (1-\lambda)x_2$, $0 < \lambda < 1$. For example in the plane a triangle has 3 extreme points, and the sphere has all its boundary points as extreme points. The following theorem gives a very important characterization of a certain kind of convex set.

Theorem (Krein-Milman) 1.3.1: A closed, bounded convex set in R^n is the convex hull of its extreme points.

A (convex) polyhedron is the intersection of finitely many half spaces. Using matrix notation we can define a polyhedron to be the set of points $P=\{x \in R^n : Ax \leq b\}$ where A is an $m \times n$ matrix and $b \in R^m$. Polyhedral sets of this form are of central importance in mathematical programming.

1.4 Convex and concave functions

If $c \in R^n$, the linear function $f : R^n \rightarrow R$ defined by $f(x) = c^T x$ is known as a linear function on R^n.

Theorem 1.4.1: Let C be a convex polyhedron in R^n. Consider the linear programming problem

$$\min_{x \in C} f(x) = c^T x. \qquad \text{(LP)}$$

If (LP) has a solution then it occurs at some vertex of C.

Proof: Let $v_1, \ldots, v_k$ be the vertices of C, and let v be the vertex such that $f(v) = \min_{1 \leq i \leq k} \{f(v_i)\}$. Since for any $x \in C$, $x = \sum_{i=1}^{k} \lambda_i v_i$, $\lambda_i \geq 0$, and $\sum_{i=1}^{k} \lambda_i = 1$, we have that $f(x) = \sum_{i=1}^{k} \lambda_i f(v_i) \geq \sum_{i=1}^{k} \lambda_i f(v) = f(v)$.

A function $f : C \subseteq R^n \rightarrow R$, where C is a convex set, is called convex if

$$f(\lambda x_1 + (1-\lambda) x_2) \leq \lambda f(x_1) + (1-\lambda) f(x_2)$$

for any $x_1, x_2 \in C$ and $0 \leq \lambda \leq 1$. The function f is called concave iff $-f$ is convex.

If the function $f(x)$ has continuous second derivatives, then the following conditions give necessary and sufficient conditions for convexity:

a) $f(y) \geq f(x)+\nabla f(x)(y-x)$ for all $x, y \in C$, or

b) The Hessian is positive semidefinite for all $x \in C$.

Theorem 1.4.2: Let $f_i : S \subseteq R^n \rightarrow R$ be convex functions. Then

1) $\sum_{i=1}^{k} \alpha_i f_i(x)$, $\alpha_i \geq 0$, is also convex

2) $\max_{1 \leq i \leq k} (f_i(x))$ is convex

3) $\max_{1 \leq i \leq k} (0, f_i(x))$ is convex.

We are concerned here with the constrained nonlinear minimization problem of the general form

$$\min_{x \in P} f(x) \qquad \text{(NP)}$$

where P is a compact convex set in R^n and $f(x)$ is a continuous function defined on P.

A point $x^* \in P$ is said to be a relative or local minimum point if $f(x^*) \leq f(x)$ for all $\| x-x^* \| \leq \varepsilon$ for some $\varepsilon>0$. We say that x^* is a global minimum point if $f(x^*) \leq f(x)$ for all $x \in P$.

When $f(x)$ is convex the problem (NP) is referred to as a convex programming problem, and when $f(x)$ is nonconvex we are talking about nonconvex programming. When the objective function $f(x)$ is convex (or more generally quasiconvex) then every local minimum is also global. This is no longer true for nonconvex functions.

Theorem 1.4.3: Suppose P is a convex compact set and $f : P \subseteq R^n \rightarrow R$ is a convex function. Then every local minimum of f over P is also global.

Proof: Let x^* be a local minimum and suppose that there exists another point y such that $f(y) < f(x^*)$. Then on the line $\lambda y+(1-\lambda)x^*$ $(0 < \lambda < 1)$ we have

$$f(\lambda y+(1-\lambda)x^*) \leq \lambda f(y)+(1-\lambda)f(x^*) < f(x^*)$$

contradicting the fact that x^* is a local minimum.

The above theorem makes convex programming a much easier problem to solve than the general nonlinear programming problem. Consider now the case where $f(x)$ is a concave function. In this case we may have many local minima which are not global. However, this problem has the following important property that also characterizes linear programming.

Theorem 1.4.4: Consider the problem

$$\text{global} \min_{x \in P} f(x)$$

where $f(x)$ is a continuous concave function defined on the bounded polyhedron P. Then every global (and local) minimum is attained at some vertex of P.

Proof: Similar to that of Theorem 1.4.1.

Note that since $\min f(x) = -\max(-f(x))$ minimization of a convex function is equivalent to maximization of a concave function(and vice versa). For continuously differentiable functions convexity and local optima are characterized using the gradient and the Hessian matrix of the function (e.g. [LUEN84], [STOE70]). For additional details regarding different convexity results see [MANG69], [GRUN67] and [ROCK70].

1.5 Convex envelopes

An important concept in nonconvex optimization is that of the convex envelope of a function.

Definition 1.5.1: Let $f: S \to R$ be a lower semi-continuous function, where S is a nonempty subset (of its domain) in R^n. Then the convex envelope of $f(x)$ taken over S is a function $F_S(x)$ such that

i) $F_S(x)$ is convex on the convex hull $Co(S)$

ii) $F_S(x) \leq f(x)$ for all $x \in S$

iii) If $h(x)$ is any convex function defined on $Co(S)$ such that $h(x) \leq f(x)$ for all $x \in S$, then $h(x) \leq F_S(x)$ for all $x \in Co(S)$.

From this definition, the convex envelope of a function is actually the best convex underestimator over S. Convex envelopes were first introduced by Kleibohm [KLEI67], who proved that with each nonconvex optimization problem is associated a convex one whose global solution is the same as that of the original problem. More precisely we have the following:

Theorem 1.5.1: Consider the problem

$$\text{global} \min_{x \in P} f(x)$$

where P is a convex set in R^n, and assume that the global minimum occurs at $x^* \in P$. Let $F(x)$ be the convex envelope of $f(x)$ over P. Then we have

$$\min_{x \in P} f(x) = \min_{x \in P} F(x)$$

and

$$\{y \in P : f(y) = \min_{x \in P} f(x)\} \subseteq \{y \in P : F(y) = \min_{x \in P} F(x)\}.$$

Proof: By definition $F(x) \leq f(x)$ for all $x \in P$. Therefore

$$\min_{x \in P} F(x) \leq \min_{x \in P} f(x) = f(x^*).$$

The constant function $G(x) = f(x^*) \leq f(x)$ is a convex underestimating function. Again by the definition of the convex envelope we have that $F(x) \geq f(x^*)$ for all x and so

$$\min_{x \in P} F(x) \geq f(x^*) = \min_{x \in P} f(x).$$

We prove the second part by contradiction. Let x^* be a global minimum of $f(x)$ over P, and suppose that x^* is not a minimum of $F(x)$ over P. Let y^* be the minimum. Then

$$F(y^*) < F(x^*) \leq f(x^*) = f^*.$$

Consider now the function $H(x) = \max(f^*, F(x))$. Then $H(x)$ is convex since the maximum of two convex functions is convex. Now $H(x) \geq F(x)$ for all $x \in P$ and

$H(y^*) > F(y^*)$, which contradicts the fact that $F(x)$, the convex envelope, is the largest convex underestimating function. Therefore $F(y^*) = F(x^*)$ and x^* globally minimizes $F(x)$ over P.

This theorem suggests that we may attempt to solve a nonconvex problem by solving the corresponding convex one where the objective function is the convex envelope of the original one. The difficulty, however, is that except in special cases, finding the convex envelope of a function is much more difficult than computing its global minimum. The next theorems describe some the of fundamental properties of convex envelopes.

Theorem 1.5.2: ([FALK76]) Let $v_1, \ldots, v_k$ be the vertices of a bounded polyhedron P. The convex envelope $F(x)$ of a concave function $f(x)$ over P can be expressed as

$$F(x) = \min_{(a_1, \ldots, a_k)} \sum_{i=1}^{k} a_i f(v_i)$$

$$\text{s.t.} \sum_{i=1}^{k} a_i v_i = x, \ \sum_{i=1}^{k} a_i = 1, \ a_i \geq 0, \ i=1, \ldots, k.$$

Proof: First we prove that $F(x)$ is convex. Let $0 \leq \lambda \leq 1$ and $x_1, x_2 \in P$ and a^1, a^2 solve $F(x_1)$ and $F(x_2)$. Then

$$F(\lambda x_1 + (1-\lambda)x_2) \leq \sum_{i=1}^{k} [\lambda a^1{}_i + (1-\lambda)a^2{}_i] f(v_i)$$

$$= \lambda \sum_{i=1}^{k} a^1{}_i f(v_i) + (1-\lambda) \sum_{i=1}^{k} a^2{}_i f(v_i) = \lambda F(x_1) + (1-\lambda) F(x_2)$$

so $F(x)$ is a convex function over P.

If $x \in P$, then $F(x) = \sum_{i=1}^{k} a_i f(v_i) \leq f(x)$ so that $F(x)$ underestimates f over the polyhedron P.

Suppose $h(x)$ is some convex function defined over P which underestimates $f(x)$ over P. Assume $F(x) < h(x)$ for some $x \in P$. Let $\bar{a}$ be the solution of the program defining $F(x)$. Then we have

$$F(x) < h(x) = h(\sum_{i=1}^{k} \bar{a}_i v_i) \le \sum_{i=1}^{k} \bar{a}_i h(v_i) \le \sum_{i=1}^{k} \bar{a}_i f(v_i) = F(x)$$

which is a contradiction.

Theorem 1.5.3: If $A \subseteq B \subseteq R^n$ are subsets in the domain of $f(x)$ then $F_A(x) \ge F_B(x)$ for all $x \in A$.

Theorem 1.5.4: Let S be a simplex generated by the vertices $v_0, v_1, \ldots, v_n$, and let $f(x)$ be a concave function defined on S. Then the convex envelope $F(x)$ of $f(x)$ over S is a linear function.

Proof: First we prove the existence of a linear function $l(x) = c^T x + b$, where $c \in R^n$ and $b \in R$, such that $l(v_i) = f(v_i)$ for $i=0, \ldots, n$.

Consider the system of $n+1$ equations with $n+1$ unknows c, b:

$$c^T v_i + b = f(v_i), \; i=0, \ldots, n$$

Subtract all equations from the first one to obtain

$$(v_0 - v_i)^T c = f(v_0) - f(v_i) = a_i, \; i=1, \ldots, n$$

or written in matrix form, $Vc = a$. The coefficient matrix V is nonsingular since the vectors $(v_0 - v_i)$, $i=1, \ldots, n$ are linearly independent. Therefore c is defined uniquely, which also gives a unique value of b.

Note that for each x in S we have that $x = \sum_{i=0}^{n} \lambda_i v_i$ where $\lambda_i \ge 0$ and $\sum_{i=0}^{n} \lambda_i = 1$. Using the fact that $f(x)$ is concave we get

$$f(x) = f(\sum_{i=0}^{n} \lambda_i v_i) \ge \sum_{i=0}^{n} \lambda_i f(v_i) = \sum_{i=0}^{n} \lambda_i l(v_i) = l(x).$$

To show that $l(x)$ is the convex envelope of $f(x)$ over S, we note that

$$F_S(x) = F_S(\sum_{i=0}^{n} \lambda_i v_i) \le \sum_{i=0}^{n} \lambda_i F_S(v_i) = \sum_{i=0}^{n} \lambda_i l(v_i) = l(x)$$

which implies that $F_S(x) = l(x)$.

Theorem 1.5.5: Let $F_1(x)$ and $F_2(x)$ be the convex envelopes of $f_1(x)$ and $f_2(x)$ on S, and let $F(x)$ be the convex envelope of $f_1(x)+f_2(x)$. Then $F(x) \geq F_1(x)+F_2(x)$.

Theorem 1.5.6: Let $S \subseteq \bigcup_{i=1}^{k} S_i$ and F_i be the convex envelope of f on S_i, $i=1, \cdots, k$. Then

$$\min_{x \in S} f(x) \geq \min_{1 \leq i \leq k} \{\min_{x \in S_i} F_i(x)\}$$

with equality holding whenever $S = \bigcup_{i=1}^{k} S_i$.

Later on we will discuss the generation of convex envelopes of concave functions using separable programming techniques [ROSE86]. More details on the use of convex envelopes can be found in [HORS76], [HORS84], [GROT85], [FALK69] and [AL-K83].

1.6 Exercises

1. Consider the linear programming problem

$$\min_{x \in P} c^T x$$

where $P = \{x \in R^n : Ax = b, x \geq 0\}$, and A is an $m \times n$ matrix. Let $S \subseteq R^m$ be the largest set such that P is nonempty for any $b \in S$. The optimal solution x^* will of course depend on the vector b, that is, $x^* = x^*(b)$. Let $\phi(b) = c^T x^*(b)$, be the corresponding optimal value of the objective function for any value $b \in S$.

i) Prove that the set S is convex.

ii) Prove that $\phi(b)$ is a convex function of b, for $b \in S$.

2. Prove that the function $f(x)$ is convex in R^n iff for any $x, y \in R^n$ the function $F(\lambda) = f(\lambda x + (1-\lambda)y)$ is convex for $0 \leq \lambda \leq 1$.

3. Let $f(x)$ be a concave function and $l(x)$ a linear function defined on the convex set $S \subseteq R^n$. If $\phi(x)$ is the convex envelope of $f(x)+l(x)$ and $F(x)$ the convex envelope of $f(x)$ prove that $\phi(x) = F(x)+l(x)$.

4. Consider the rectangle $R=\{x \in R^n : a_i \le x_i \le b_i, i=1, \ldots, n\}$ and let $f(x) = \sum_{i=1}^{n} f_i(x_i)$ be a (separable) concave function defined on R. Prove that the convex envelope of $f(x)$ is equal to the sum of the convex envelopes of the functions $f_i:[a_i, b_i] \to R$, $i=1, \ldots, n$.

5. Let $f(x, y) = xy$ be defined on the rectangle

$$R = \{(x, y): l \le x \le L, m \le y \le M\}$$

Prove that the convex envelope of $f(x, y)$ is given by

$$F(x, y) = \max\{mx+ly-lm, Mx+Ly-LM\}$$

6. Consider the following problem

$$\min f(x) = \sum_{i=1}^{n} c_i |x_i|$$

$$\text{s.t. } Ax \le 0$$

i) If $c_i \ge 0$, $i=1, \ldots, n$, prove that $f(x)$ is a piecewise linear convex function. Then reduce the above problem to a linear programming problem.
ii) If for some i, $c_i < 0$, then this problem becomes nonconvex. In this case reduce this problem to a mixed 0–1 integer program with a linear objective function.

1.7 References

[AL-K83] Al-Khayyal, F.A. and Falk, J.E. *Jointly constrained biconvex programming.* Math. Oper. Res. 8 (1983), 273-286.

[FALK69] Falk, J.E. and Soland, R.M. *An algorithm for separable nonconvex programming problems*. Manag. Sci. 15 (196), 550-569.

[FALK76] Falk, J.E. and Hoffmann, K.L. *A successive underestimating method for concave minimization problems*. Math. Oper. Res. 1 (1976), 251-259.

[GROT85] Grotzinger, S.J. *Supports and convex envelopes*. Math. Progr. 31 (1985), 339-347.

[GRUN67] Grunbaum, B. *Convex polytopes*. Wiley, N.Y. (1967).

[HORS84] Horst, R. *On the convexification of nonlinear programming problems: An applications oriented survey*. European Journ. of Oper. Res. 15 (1984), 382-392.

[HORS76] Horst, R. *An algorithm for nonconvex programming problems*. Math. Progr. 10 (1976), 312-321.

[KLEI67] Kleibohm, K. *Bemerkungen zum problem der nichtkonvex programmierung. (Remarks on nonconvex programming problems)*. Unternehmensf. 11 (1967),49-60.

[LUEN84] Luenberger, D.G. *Linear and nonlinear programming*. (2nd ed. 1984) Addison Wesley.

[MANG69] Mangasarian, O.L. *Nonlinear Programming*. McGraw-Hill Inc. N.Y. (1969).

[ROCK70] Rockafellar, R.T. *Convex Analysis*. Princeton University Press (1970).

[ROSE86] Rosen, J.B. and Pardalos, P.M. *Global minimization of large-scale concave quadratic problems by separable programming*. Math. Progr. 34 (1986), 163-174.

[STOE70] Stoer, J. and Witzgall, C. *Convexity and optimization in finite dimensions*. Springer-Verlag (1970).

Chapter 2 Optimality conditions in nonlinear programming

In this section we give a brief and broad overview of some results from nonlinear programming. It is well known that necessary conditions for local optimality can be obtained from the Kuhn-Tucker theorems. Furthermore, under suitable convexity assumptions about the objective function and the feasible domain, these conditions are sufficient. However, in the nonconvex case sufficiency is no longer guaranteed.

First we consider the Kuhn-Tucker (KT) conditions for the nonlinear problem with inequality constraints,

$$\min_{x \in S} f(x) \tag{1}$$

where

$$S = \{x : g_i(x) \le 0,\ i=1, \ldots, p\} \subseteq R^n .$$

Assume that the feasible domain S is compact and $f \in C^2$. Let x^* be a local minimum and let $J(x^*) = \{i : g_i(x^*) = 0\}$. Also assume that the vectors $\nabla g_i(x^*)$, $i \in J(x^*)$ (the gradients of the active constraints) are linearly independent.

Theorem 2.1: If x^* solves problem (1) then the following KT conditions hold for x^*.

a) $g_i(x^*) \le 0,\ i=1, \ldots, p$

b) there exist $\lambda_i \ge 0$ such that $\lambda_i g_i(x^*) = 0,\ i=1, \ldots, p$ and

c) $\nabla f(x^*) + \sum_{i=1}^{p} \lambda_i \nabla g_i(x^*) = 0$

The last two conditions are equivalent to:

d) $\nabla f(x^*) + \sum_{i \in J(x^*)} \lambda_i \nabla g_i(x^*) = 0$, and $\lambda_i \ge 0,\ i=1, \ldots, p$.

The function $L(x, \lambda) = f(x) + \sum_{i=1}^{p} \lambda_i g_i(x)$ is called the Lagrangian function.

Theorem 2.2: The KT conditions are sufficient for a constrained global minimum at x^* provided that the functions $f(x)$ and $g_i(x)$, $i=1, \ldots, p$ are convex.

Proof: By convexity of $f(x)$ and $g_i(x)$ we have

$$f(x) \geq f(x^*)+(x-x^*)^T \nabla f(x^*)$$

$$g_i(x) \geq g_i(x^*)+(x-x^*)^T \nabla g_i(x^*)$$

Multiplying the last inequalities by λ_i and adding to the first one we obtain:

$$f(x)+\sum_{i=1}^{p}\lambda_i g_i(x) \geq f(x^*)+\sum_{i=1}^{p}\lambda_i g_i(x^*)+(x-x^*)^T[\nabla f(x^*)+\sum_{i=1}^{p}\lambda_i \nabla g_i(x^*)]$$

which implies $f(x) \geq f(x^*)-\sum_{i=1}^{p}\lambda_i g_i(x) \geq f(x^*)$ for all $x \in S$, that is x^* is a global minimum.

Consider now the KT conditions for the problem with equality constraints

$$\min_{x \in S} f(x)$$

where $S=\{x : h_i(x) = 0,\ i=1, \ldots, t\} \subseteq R^n$ is compact and $f, h_i \in C^2$. Again here we assume that if x^* is a minimum, the vectors $\nabla h_i(x^*)$, $i=1, \ldots, t$ are linearly independent.

Theorem 2.3: The following KT conditions are necessary for x^* to solve the above problem:

a) $h_i(x^*) = 0$, $i=1, \ldots, t$

b) there exists μ such that $\nabla f(x^*)+\sum_{i=1}^{t}\mu_i \nabla h_i(x^*) = 0$.

Note that in the special case where all h_i are linear, the set S is a convex polytope.

Next we consider the situation when Kuhn-Tucker theory is applied to nonconvex programming. We illustrate some difficulties with the following simple examples of separable concave programming.

Example 2.1: Consider the problem

$$\min -(x_1^2+x_2^2)$$

$$\text{s.t. } x_1 \leq 1$$

The KT conditions for this problem are

$$x_1^* - 1 \leq 0,\ \lambda_1(x_1^* - 1) = 0,\ \lambda_1 \geq 0,\ \lambda_1 - 2(x_1^*) = 0,\ -2x_2^* = 0.$$

It easy to see that the KT conditions are satisfied at $x^* = (0,0)$ with $\lambda_1 = 0$ and at $x^* = (1,0)$ with $\lambda_1 = 2$. The first is a global maximum. The second is neither a local minimum nor a local maximum. The problem has no local minima; in fact, since S is unbounded, the minimum of f is unbounded.

Example 2.2:

$$\min 2x - x^2$$

$$\text{s.t. } 0 \leq x \leq 3$$

The KT conditions for this problem are

$$\lambda_1(x^* - 3) = 0,\ \lambda_2 x^* = 0,\ 2(1 - x^*) + \lambda_1 - \lambda_2 = 0,\ \lambda_1 \geq 0,\ \lambda_2 \geq 0$$

Since the objective function is concave local minima occur at the endpoints of the interval [0, 3]. The point $x^* = 3$ is the global minimum. The endpoints satisfy the KT conditions ($x^* = 0$ with $\lambda_1 = 0$, $\lambda_2 = 2$ and $x^* = 3$ with $\lambda_1 = 4$, $\lambda_2 = 0$). However we can easily see that the KT conditions are also satisfied at $x^* = 1$ (with $\lambda_1 = \lambda_2 = 0$) and that this is a global maximum point.

These examples show that for nonconvex functions KT points may not be local minima. One could instead try to determine all possible solutions to the KT conditions and then take the one with the smallest function value. The next example will show that this approach may be worse than complete enumeration.

Example 2.3:

$$\min -\sum_{i=1}^{n}(c_i x_i + x_i^2)$$

$$\text{s.t. } -1 \le x_i \le 1,\ i=1,\ldots,n$$

where the c_i, $i=1,\ldots,n$ are small positive numbers.

The unique global minimum of this problem is $-(n+\sum_{i=1}^{n} c_i)$ and occurs at the vertex $v = (1, 1, \ldots, 1)$. For this problem we can prove that we have 3^n Kuhn-Tucker points, 2^n local optima (all vertices of the hypercube), and a large number, $3^n-(2^n+1)$ of saddle points (useless KT points). Therefore global optimization by KT theory may be very inefficient.

2.1 Quadratic problems solvable in polynomially bounded time

Recently, several methods have been proposed for the solution of convex quadratic problems whose computational complexity can be shown to be bounded by a polynomial in the size of the problem data. Consider the convex quadratic programming problem:

$$\min\ c^T x + \frac{1}{2} x^T Q x \tag{1}$$

$$\text{s.t. } Ax \le b$$

where $Q_{n\times n}$ is a symmetric positive definite matrix, A is an $m\times n$ matrix and $c, x \in R^n$, $b \in R^m$. The Kuhn-Tucker conditions for this problem may be stated as the following system of equalities and inequalities.

$$Ax \le b,\ \lambda^T(Ax-b)=0 \tag{KTC}$$

$$c + Qx + A^T\lambda = 0,\ \ \lambda \ge 0.$$

In the special case where Q is a diagonal positive semidefinite matrix and the constraints have the simple form

$$\sum_{i=1}^{n} x_i = b,\ 0 \le x_i \le \beta_i,\ i=1,\ldots,n$$

a polynomially bounded algorithm was proposed in [HELG79] using an appropriate manipulation of the corresponding Kuhn-Tucker conditions. An optimal $O(n)$

algorithm using binary search to solve the Kuhn-Tucker system can be found in [PARD87A].

For the general case we observe that for any (x, λ), a solution of the (KTC), there is one that satisfies the vertices of the related polyhedron

$$S = \{(x, \lambda): c+Qx+A^T\lambda = 0, \lambda \geq 0\} \subseteq R^{n+m}.$$

Therefore, it is enough to consider the equalities and inequalities that define S. Assume that all problem inputs are integers. Let L be the length of the input to problem (1), i.e. the number of symbols 0, 1 that are necessary to write (1) in the binary number system (binary encoding). Then $L = L_1+L_2$ where

$$L_1 = [\sum_{i,j=1}^{n} \log_2(|q_{ij}|+1)+\sum_{i=1}^{n}\log_2(|c_i|+1)]$$

$$L_2 = [\sum_{i,j=1}^{mn} \log_2(|a_{ij}|+1)+\sum_{i=1}^{n}\log_2(|b_i|+1)+\log_2 mn+1].$$

L. Khachian's ellipsoid algorithm [KHAC79] for linear inequalities decides the consistency of the system $Ax \leq b$ in polynomially bounded time. This algorithm has complexity $O(n^3(n^2+m)L_2)$ (and $O(nL_2)$ space complexity). In a paper by Kozlov, Tarasov and Khachian [KOZL79], the ellipsoid algorithm is applied to decide the consistency of the system

$$\lambda \geq 0,\ c+Qx+A^T\lambda = 0 \quad (x, \lambda) \in R^{m+n}$$

and an $O((n+m)^5L)$ algorithm is given. Boundedness below of the convex function $f(x)$ on the feasible set is equivalent to consistency of the above system. Since the binary encoding of this system has size at most $2L$ (L is the size of the input for the quadratic problem), it can be shown that the optimal function value is of the form $f^* = s/t$, where t and s are relatively prime integers, with $|t| \leq 2^{5L}$ and $|s| \leq 2^{4L}$. But it is enough to approximate f^* with a real number r that satisfies

$$|f^* - r| < \frac{1}{2(2^{4L})^2} = 2^{-8L-1}$$

In order to find the solution $f^* = t/s$, [KOZL79], it is enough to determine the consistency of a sequence of $13L+2$ systems P_k of the form

$$Ax \leq b$$

$$f(x) \leq t_k / s_k$$

Again, a version of the ellipsoid algorithm is applied for these systems. Each problem P_k is solved by the ellipsoid algorithm (decision version) in $O(n^2 L)$ iterations. For details see [KOZL79], [AKGU84] and [CHUN81].

As a final remark we should mention here that computational studies have indicated that the ellipsoid methods are far from being competitive in practice with other existing methods for convex quadratic programming. However, when the ellipsoid algorithm that was first proposed by Shor in [SHOR70] is applied for solving the nonlinear programming problem

$$\min_{x \in S} f(x) \text{ where } S = \{x \in R^n : f_i(x) \leq 0,\ i=1, \ldots, m\}$$

where each f_i is convex, it is has been observed to have some remarkable properties. For example, regarding robustness, other nonlinear programming algorithms are far more sensitive to the starting point than the ellipsoid algorithm is to the starting ellipsoid. It has also been observed in computational experiments that on problems with several local minima, the ellipsoid algorithm often converges to a global optimum. For details on such computational experiments see [ECKE85].

A polynomially bounded algorithm which has been proposed by [KAPO86] for convex quadratic programs, is faster than the ellipsoid method and is based on Karmarkar's method for linear programming [KARM84].

Recently a new iterative algorithm for convex quadratic programming has been developed by Ye and Tse [YE86], based on the iterated application of the cutting objective technique followed by optimization over an interior ellipsoid centered at the current solution to create a sequence of points that converge to the solution. The number of iterations of this new algorithm is bounded by $O(nL)$, and each iteration requires $O(n^3 L^2)$ arithmetic operations, where n is the dimension of the problem and L the input size.

In the case of nonconvex programming the general problem is NP-hard. However, certain classes of problems have been identified that can be solved in polynomially bounded time. Consider the problem of minimizing a quadratic function over the unit hypercube, that is

$$\min_x c^T x + \frac{1}{2} x^T Q x$$

$$\text{s.t. } 0 \le x_i \le 1, \; i=1, \ldots, n$$

where Q is a symmetric negative semidefinite matrix. In the special case where the matrix Q satisfies $q_{ij} \le 0$ this problem can be solved as a minimum cut network problem in polynomially bounded time (see [PICA82]). More recently, a linear time algorithm was given for the same problem, when the graph defined by Q is series-parallel. For details concerning the last algorithm see [BARA86].

2.2 Some general remarks

In certain classes of nonlinear problems the local solution is always the global one. For example, as we proved earlier, in minimization problems with a convex objective function subject to convex constraints, the local minimizer is the global solution. In this case the problem is easier since we need only find a local solution.

It seems to be a very difficult problem to characterize a general class of constrained problems which have a unique local minimum (and therefore an easily found global minimum), since such characterizations rely on properties of the objective function and the feasible domain. For a number of results related to this class of problems see [MANG69], [HORS82], [AVRI76], [ZANG76], [GASA85] and [GASA84].

We close this chapter with a final remark about local minima in nonconvex programming. In [PARD87B], it is proved that the problem of checking local optimality for a feasible point and the problem of checking if a local minimum is strict, are NP-hard problems (even in the case of a separable indefinite quadratic objective function with network constraints). As a consequence, the problem of checking whether a function is locally strictly convex is also NP-hard. Related results on the complexity of checking local optimality for nonconvex programming problems were obtained by [MURT85] and [VERG86].

2.3 Exercises

1. Consider the linearly constrained quadratic problem

$$\min_{x \in P} f(x) = c^T x + \frac{1}{2} x^T Q x$$

where $P = \{x \in R^n : A^T x = b\}$, A is an $n \times t$ matrix with $t \le n$, $rank(A) = t$, and Q is a symmetric positive definite matrix. Prove that this problem has a unique Kuhn-Tucker point (x^*, λ^*), where λ^* is determined by the system of linear equations:

$$(A^T Q^{-1} A)\lambda^* = -(b + A^T Q^{-1} c).$$

2. Consider the concave minimization problem

$$\min -x^2 - y^2$$

$$\text{s.t. } x + 4y \le 5,\ x \le 1,\ x, y \ge 0.$$

Find the local and global minima of this problem. If you consider the KT points for this problem what can you say about the optimal solution?

3. Given the following indefinite quadratic programming problem

$$\min_x \frac{1}{2} x_1^2 - \frac{1}{2} x_2^2 - \varepsilon x_1$$

$$\text{s.t. } -x_1 + 2x_2 \le 0$$

$$-x_1 - 2x_2 \le 0$$

prove that:
i) If $\varepsilon = 0$ this problem has a unique minimum at $x = (0,0)$.
ii) If $\varepsilon > 0$ it has local minima at the points $x = \frac{2}{3}\varepsilon(2, \pm 1)$ and a saddle point at $(\varepsilon, 0)$.

4. Consider the following quadratic problem (No convexity is assumed):

$$\min c^T x + \frac{1}{2} x^T Q x$$

$$\text{s.t. } Ax = b.$$

Prove that x^* is a local minimum if and only if it is a global minimum.

5. Let $P(\alpha, \beta)$ be a family of convex polytopes defined by

$$P(\alpha, \beta) = \{x : A(\alpha)x \geq b(\beta)\} \subseteq R^4$$

where

$$A(\alpha) = \begin{bmatrix} 0 & -1 & 1 & 0 \\ 0 & 1 & 1 & 0 \\ -1 & 0 & 1 & 0 \\ 1 & 0 & 1 & \alpha \end{bmatrix}, \quad b(\beta) = \begin{bmatrix} 1 \\ 1 \\ 1 \\ 1+\beta \end{bmatrix}.$$

Consider the convex quadratic problem

$$\min_{x \in P(\alpha,\beta)} \sum_{i=1}^{4} x_i^2$$

Find the values of the parameters α, β for which the above problem has a unique solution and compute that solution.

2.4 References

[AKGU84] Akgul, M. *Topics in relaxation and ellipsoidal methods.* Lecture Notes in Math., 97 (1984), Pitman Publ. Co.

[AVRI76] Avriel, M. *Nonlinear Programming: Analysis and Methods.* Prentice Hall Inc. Englewood Cliffs, N.J. (1976).

[BARA86] Barahona, F. *A solvable case for quadratic 0-1 programming.* Discrete Appl. Math. 13 (1986), 23-26.

[BURD77] Burdet, C.A. *Elements of a theory in non-convex programming.* Naval Res. Log. Quarterly Vol. 24, No. 1 (1977), 47-66.

[CHUN81] Chung, S.J. and Murty, K.G. *A polynomially bounded ellipsoid algorithm for convex quadratic programming.* Nonlinear Programming 4 (ed. O.L. Mangasarian, R.R. Meyer and S.M. Robinson), Acad. Press, (1981), 439-485.

[ECKE85] Ecker, J.G. and Kupferschmid, M. *A computational comparison of the ellipsoid algorithm with several nonlinear programming algorithms.* SIAM J. Control and Optimization, Vol. 23, No. 5 (1985), 657-674.

[FALK67] Falk, J. *Lagrange multipliers and nonlinear programming.* Journ. of Math. Anal. and Applic. 19 (1967), 141-159.

[FALK69] Falk, J. *Lagrange multipliers and nonconvex programs.* SIAM J. on Control and Optimization 7 (1969), 534-545.

[GASA84] Gasanov, I.I. and Rikun, A.D. *On necessary and sufficient conditions for uniextremality in nonconvex mathematical programming problems.* Soviet Math. Dokl. Vol 30, No. 2 (1984), 457-559.

[GASA85] Gasanov, I.I. and Rikun, A.D. *The necessary and sufficient conditions for single extremality in nonconvex problems of mathematical programming.* USSR Comput. Maths. Math. Phys., Vol. 25, No. 3 (1985), 105-113.

[HELG80] Helgason, R., Kennington, J., and Lall, H. *A polynomially bounded algorithm for a singly constrained quadratic program.* Math. Progr. 18 (1980), 338-343.

[HORS82] Horst, R. *A note on functions whose local minima are global.* JOTA 36 (1982), 457-463.

[KAPO86] Kapoor, S. and Vaidya, P.M. *Fast algorithms for convex quadratic programming and multicommodity flows.* Working paper, Comp. Sc. Dept. (1986), University of Illinois at Urbana-Champaign.

[KARM84] Karmarkar, N. *A new polynomial-time algorithm for linear programming.* Combinatorica 4 (1984), 373-395.

[KHAC79] Khachian, L.G. *A polynomial algorithm for linear programming.* Soviet Math. Dokl. Vol. 20 (1979), 191-194.

[KOZL79] Kozlov, M.K., Tarasov, S.P., and Khachian, L.G. *Polynomial solvability of convex quadratic programming*. Soviet Math. Dokl. Vol. 20 (1979), 1108-1111.

[LUEN84] Luenberger, D.G. *Linear and Nonlinear Programming*. Addison Wesley (1984).

[MANG69] Mangasarian, O.L. *Nonlinear Programming*. McGraw-Hill, N.Y. (1969).

[MURT85] Murty, K.G. and Kabadi, S.N. *Some NP-complete problems in quadratic and nonlinear programming*. Technical report 85-23 (1985), Dept. of Industr. and Oper. Eng., Univ. of Michigan.

[PARD87A] Pardalos, P.M. and Kovoor, N. *An algorithm for singly constrained quadratic programs*. Technical Report CS-87-06 (1987), Computer Sci. Dept., The Penn. State Univ.

[PARD87B] Pardalos, P.M. and Schnitger, G. *Checking local optimality in constrained quadratic programming is NP-hard*. Technical Report (1987), Computer Sci. Dept., The Penn. State Univ.

[PICA82] Picard, J.C. and Queyranne, M. *Selected applications of min cut in networks*. INFOR Vol. 20, No. 4 (1982), 395-422.

[SHOR70] Shor, N.Z. *Convergence rate of the gradient descent method with dilatation of the space*. Cybernetics 6 (1970), 102-108.

[VERG86] Vergis, A., Steiglitz, K., and Dickinson, B. *The complexity of analog computation*. Math. and Computers in Simulation 28 (1986), 91-113.

[YE86] Ye, Y. and Edison, T. *A polynomial time algorithm for convex quadratic prgramming*. Working paper (1986), Depart. of Engin.-Econom. Systems, Stanford Univ.

[ZANG75] Zang, I. and Avriel M. *On functions whose local minima are global*. JOTA 16 (1975), 183-190.

[ZANG76] Zang, I. *A note on functions whose local minima are global*. JOTA 18 (1976), 556-559.

Chapter 3 Combinatorial optimization problems that can be formulated as nonconvex quadratic problems

A broad class of difficult combinatorial optimization problems can be formulated as nonconvex quadratic global minimization problems. In this chapter we discuss some of these problems and give the equivalent formulations.

3.1 Integer Programming

A number of important problems in operations research, graph theory, and mathematical programming are formulated as zero-one linear problems

$$\min_{x} c^T x$$

$$\text{s.t. } Ax \le b,\ x_i = 0,\ 1\ (i=1, \ldots, n)$$

where A is an $m \times n$ matrix, $c \in R^n$ and $b \in R^m$.

It was shown in [RAGH69] that the zero-one integer programming problem is equivalent to a concave quadratic minimization problem subject to linear constraints. Let $e^T=(1, \ldots, 1) \in R^n$ denote the vector whose components are all equal to 1. Then the zero-one integer linear programming problem is equivalent to the following concave minimization problem:

$$\min f(x) = c^T x + \mu x^T (e-x)$$

$$\text{s.t. } Ax \le b,\ 0 \le x \le e$$

where μ is a sufficiently large positive number.

The function $f(x)$ is concave since $-x^T x$ is concave. The equivalence of the two problems is based on the fact that a concave function attains its minimum at a vertex and $x^T(x-e) = 0$, $0 \le x \le e$ implies that $x_i = 0$ or 1, for $i=1, \ldots, n$. Note that a vertex of the feasible domain is not necessarily a vertex of the hypercube, but the global minimum is attained only when $x^T(e-x) = 0$, provided that μ is large enough.

More generally [GIAN76], a nonlinear nonconvex integer program can be reduced to an equivalent concave minimization program (assuming that the nonconvex objective function is bounded and satisfies the Lipschitz condition). In this approach a penalty μ is also used. Some estimates on the size of the penalty μ are given in [GIAN76] and [KALA82].

3.2 Quadratic 0–1 programming

The general quadratic zero-one problem has the following form:

$$\min_x f(x) = c^T x + x^T Q x$$

$$\text{s.t. } x_i = 0, 1, \; i=1, \ldots, n$$

where Q is an $n \times n$ symmetric matrix. Problems of this form are encountered in many diverse fields, such as computer aided layout, portfolio selection and site selection for electric message systems (for references see [GULA81]). Since NP-hard problems, for example the clique problem, can be formulated as a quadratic zero-one problem, the problem is itself NP-hard.

Given any real number μ, let $\overline{Q} = Q + \mu I$ and $\overline{c} = c - \mu e$. Then the above zero-one quadratic problem is equivalent to

$$\min_x \overline{f}(x) = \overline{c}^T x + x^T \overline{Q} x$$

$$\text{s.t. } x_i = 0, 1 \; i=1, \ldots, n$$

If we choose μ such that $\overline{Q} = Q + \mu I$ becomes a negative semidefinite matrix (e.g. $\mu = -\lambda$, where λ is the largest eigenvalue of Q), then the objective function $\overline{f}(x)$ becomes concave and the constraints can be replaced by $0 \leq x \leq e$. Therefore the problem is equivalent to the minimization of a quadratic concave function over the unit hypercube.

3.3 Quadratic assignment problem

The general quadratic assignment problem (QAP) can be stated as

$$\min_x x^T S x$$

$$\text{s.t. } x \in X = \{x : \sum_{i=1}^{n} x_{ij} = 1,\ j=1, \ldots, n,\ \sum_{j=1}^{n} x_{ij} = 1,\ i=1, \ldots, n,\ x_{ij} \in \{0,1\}\}$$

Here $x \in R^{n^2}$ and the matrix S has nonnegative entries. Next we are going to formulate (QAP) as a concave minimization problem. First we state a useful lemma [BAZA82]:

Lemma 3.3.1: Let $Q = S - \alpha I$ where $\alpha = 1 + \| S \|_\infty$. Then the matrix Q is negative definite.

Proof: Let $m = n^2$ and $x = (x_{11}, \ldots, x_{nn}) = (x_1, \ldots, x_m)$. The we have

$$x^T Q x = \sum_{i=1}^{m} q_{ii} x_i^2 + 2 \sum_{i=1}^{m-1} \sum_{j=i+1}^{m} q_{ij} x_i x_j$$

$$= \sum_{i=1}^{m} (q_{ii} + \sum_{\substack{j=1 \\ j \neq i}}^{m} q_{ij}) x_i^2 - \sum_{i=1}^{m-1} \sum_{j=i+1}^{m} q_{ij} (x_i - x_j)^2$$

$$= \sum_{i=1}^{m} (-\alpha + \sum_{j=1}^{m} s_{ij}) x_i^2 - \sum_{i=1}^{m-1} \sum_{j=i+1}^{m} s_{ij} (x_i - x_j)^2 \leq \sum_{i=1}^{m} (-\alpha + \sum_{j=1}^{m} s_{ij}) x_i^2 .$$

Therefore $x^T Q x < 0$ for any $x \neq 0$ and the proof is complete.

Theorem 3.3.1: The (QAP) is equivalent to the following concave programming problem:

$$\min_x x^T Q x$$

$$\text{s.t. } x \in \Omega = \{x : \sum_{i=1}^{n} x_{ij} = 1,\ j=1, \ldots, n,\ \sum_{j=1}^{n} x_{ij} = 1,\ i=1, \ldots, n,\ x \geq 0\}$$

Proof: Note that $x^T \alpha I x = \alpha \sum_{i=1}^{n} \sum_{j=1}^{n} x_{ij} = \alpha n$ which is a constant. Now the feasible sets X and Ω have the same extreme points. Since the concave function achieves its minimum at some extreme point the proof is complete.

3.4 The 3-dimensional assignment problem

Frieze [FRIE74] has reduced the 3-dimensional assignment problem (3-AP) to a bilinear programming problem and then to a special convex maximization problem. The (3-AP) has the form:

$$\max \sum_{i=1}^{m}\sum_{j=1}^{n}\sum_{k=1}^{p} a_{ijk}x_{ijk}$$

$$\text{s.t.} \quad \sum_{j=1}^{n}\sum_{k=1}^{p} x_{ijk} = 1,\ i=1,\ldots,m$$

$$\sum_{i=1}^{m}\sum_{k=1}^{p} x_{ijk} \le b_j,\ j=1,\ldots,n$$

$$\sum_{i=1}^{m}\sum_{j=1}^{n} x_{ijk} \le c_k,\ k=1,\ldots,p$$

$$x_{ijk} \in \{0,\ 1\}$$

where we may assume without loss of generality that $b_j > 0$ for $j=1,\ldots,n$, $c_k > 0$ for $k=1,\ldots,p$ and $a_{ijk} \ge 0$.

Theorem 3.4.1: There is a one to one correspondence between the solutions to (3-AP) and the problem below:

$$\max \phi(y,z) = \sum_{i=1}^{m}\sum_{j=1}^{n}\sum_{k=1}^{p} a_{ijk}y_{ij}z_{ik}$$

$$\text{s.t.} \quad \sum_{j=1}^{n} y_{ij} = 1, \quad \sum_{k=1}^{p} z_{ik}=1, \quad i=1,\ldots,m$$

$$\sum_{i=1}^{m} y_{ij} \le b_j, \quad j=1,\ldots,n, \quad \sum_{i=1}^{m} z_{ik} \le c_k, \quad k=1,\ldots,p$$

$$y_{ij} \ge 0, \quad z_{ik} \ge 0.$$

Therefore the original (3-AP) has been reduced to an equivalent bilinear problem with 0–1 solutions. Note now that the objective function in the bilinear formulation can be replaced by

$$\phi(y,z) = \frac{1}{2}\sum_{i=1}^{m}\sum_{j=1}^{n}\sum_{k=1}^{p} a_{ijk}(y_{ij}+z_{ik})(y_{ij}+z_{ik}-1)$$

Since $a_{ijk} \geq 0$, this is a convex function, and the original (3-AP) is equivalent to a convex maximization problem.

3.5 Bilinear Programming

Consider the following bilinear programming problem,

$$\max f(x,y) = c^T x + c^T y + x^T Q y \qquad \text{(BLP)}$$

$$\text{s.t. } Ax = b,\ Ay = b,\ x, y \geq 0$$

where Q is an $n \times n$ symmetric positive definite matrix, $c, x, y \in R^n$ and $b \in R^m$. Konno [KONN76], proved that this problem is equivalent to the convex maximization problem,

$$\max \phi(x) = 2c^T x + x^T Q x$$

$$\text{s.t. } Ax = b,\ x \geq 0.$$

If x^* is the solution to the above problem, then (x^*, x^*) is the optimal solution to (BLP). Conversely if (x^*, y^*) is optimal to (BLP), then both x^* and y^* are optimal for the second problem. For some more general results see also [THIE80].

3.6 Linear Complementarity Problems

In its standard notation the Linear Complementarity Problem (LCP) can be stated as:

Given $M \in R^{n \times n}$ and $q \in R^n$,

find $x \in R^n$ such that $x \geq 0$, $Mx+q \geq 0$ and $x^T(Mx+q) = 0$

(or prove that such an x does not exist).

The LCP unifies a number of problems of operations research and in particular it contains the linear programming (primal-dual) problem and also the convex quadratic programming problem as special cases (see [COTT68] where the LCP is called the "Fundamental Problem").

Let $S=\{x: Mx+q \geq 0, x \geq 0\}$ be the corresponding feasible domain. Mangasarian [MANG78], proved the following interesting result:

Theorem 3.6.1: The (LCP) is equivalent to the following (piecewise linear) concave minimization problem

$$\min_{x \in S} f(x) = \sum_{i=1}^{n} \{\min(0, M_i^T x - x_i + q_i) + x_i\}$$

where M_i^T is the ith row of M, and q_i the ith component of q.

Proof: It is clear that $Mx+q \geq 0$, $x \geq 0$ iff $\min(0, M_i^T x + q_i) + x_i \geq 0$ for all $i=1, \ldots, n$. Therefore $f(x) \geq 0$ for all $x \in S$. Now, v solves the (LCP) iff $v \in S$ and $v_i = 0$ or $M_i^T v - x_i + q_i = 0$, $i=1, \ldots, n$. This is equivalent to $f(v) = 0$.

The function $f(x)$ is concave since $\min(0, M_i^T x - x_i + q_i)$ is concave (being the min of two linear functions). Hence, a useful property of (LCP) problems is that if they have a finite solution then they have a solution that occurs at some vertex of S.

The linear complementarity problem we defined above is a special case of a more general problem:

Given the matrices $A, B \in R^{m \times n}$, $C \in R^{m \times p}$ and a vector $q \in R^m$, find $x, y \in R^n$, $z \in R^p$ satisfying

$$Ax+By+Cz = q, \quad x^T y = 0 \quad x, y, z \geq 0 \qquad \text{(GCP)}$$

Consider the piecewise linear concave function $f(x,y,z) = \sum_{i=1}^{n} \min(x_i, y_i)$.Then the following is true [THIE80]:

Theorem 3.6.2: (x^*, y^*, z^*) is a solution of (GCP) if and only if (x^*, y^*, z^*) is an optimal solution of the following concave problem

$$\min f(x,y,z) = \sum_{i=1}^{n} \min(x_i, y_i)$$

$$\text{s.t. } Ax+By+Cz = q, \ x, y, z \geq 0$$

with $f(x^*, y^*, z^*) = 0$.

3.7 Max-min problems

The max-min problem can be stated as

$$\max_x \min_y \{c^T x + d^T y\}$$

$$\text{s.t } Ax+By \leq b, \ x, y \geq 0$$

where $x, c \in R^s$, $y, d \in R^t$, A is an $m \times s$ matrix, B is an $m \times t$ matrix and $b \in R^m$.

Falk [FALK73] proved that this problem is equivalent to the problem of maximizing a convex function subject to linear constraints (a global concave minimization problem). As a consequence of that result, there is a vertex (x^*, y^*) of the constraint set that solves the problem. For a different approach see also [IVAN76].

3.8 Exercises

1. Consider the following Knapsack problem: Find a feasible solution to the system

$$\sum_{i=1}^{n} a_i x_i = b, \ x_i = 0,1, \ i=1, \ldots, n \qquad \text{(KFP)}$$

where a_i, $i=1, \ldots, n$ and b are positive integers. Formulate the (KFP) as a linear complementarity problem, with $M_{(n+2)\times(n+2)}$ a negative definite triangular matrix, that is, the (KFP) can be reduced to a quadratic concave minimization problem.

2. Formulate the linear complementarity problem $LCP(M,q)$ as a mixed zero-one integer linear program.
Hint: Consider the following mixed zero-one integer linear program:

$$\max \alpha$$

$$\text{s.t. } 0 \le (My)_i + \alpha q_i \le 1 - z_i,\ i=1, \ldots, n$$

$$0 \le y_i \le z_i,\ z_i \in \{0,1\},\ i=1, \ldots, n$$

3. Prove that the following separable problem

$$\max_{x \in P} x_1^2 + \cdots + x_n^2$$

where P is a polyhedral set in R^n, is NP-complete.

4. Consider the following integer quadratic problem:

$$\min_{x \in D} f(x) = c^T x + \frac{1}{2} x^T M x$$

where $D = \{x : Ax \le b,\ x_i \in \{-1,1\},\ i=1, \ldots, n\}$, M is an $n \times n$ symmetric matrix, $c \in R^n$, $b \in R^m$ and A is an $m \times n$ matrix.
a). Prove that every quadratic zero-one program can be reduced to an equivalent one of the above form.
b). Suppose that the feasible domain D and the symmetric matrix M are given. Let $x_0 \in \{\pm 1\}^n$ be a feasible integer point. If $c = -(M+\varepsilon I)x_0$, where the matrix $M+\varepsilon I$ is positive semidefinite for some $\varepsilon > 0$, then prove that $\min_{x \in D} f(x) = f(x_0)$ (see also [PARD86]).

5. Consider the following Subset-Sum Problem (SSP), which is known to be NP-complete.
Given positive integers $k_0; k_1, \ldots, k_n$, check if there is a solution to

$$\sum_{i=1}^{n} k_i x_i = k_0,\ x_i \in \{0,1\},\ i=1, \ldots, n.$$

Formulate the above (SSP) problem as a global optimization problem with an indefinite quadratic objective function.

3.9 References

[BAZA82] Bazaraa, M.S. and Sherali, H.D. *On the use of exact and heuristic cutting plane methods for the quadratic assignment problem.* J. Oper. Res. Soc. 33 (1982), 991-1003.

[COTT68] Cottle, R.W. and Dantzig, G.B. *Complementarity pivot theory of mathematical programming.* In: G.B. Dantzig and A.F. Veinott, Jr., eds. Mathematics of the Decision Sciences, Part 1 (Amer. Math. Society, Providence, RI, 1968), 115-136.

[FALK73] Falk, J.E. *A linear max-min problem.* Math. Progr. 5 (1973), 169-188.

[FRIE74] Frieze, A.M. *A bilinear programming formulation of the 3-dimensional assignment problem.* Math. Progr. 7 (1974), 376-379.

[GIAN76] Giannessi, F. and Niccolucci, F. *Connections between nonlinear and integer programming problems.* In: Symposia Mathematica Vol. XIX, Inst. Nazionale Di Alta Math. Academic Press (1976), 161-176.

[GULA81] Gulati, V.P., Gupta, S.K., and Mittal, A.K. *Unconstrained quadratic bivalent programming problem.* European J. of Oper. Res. 15 (1981), 121-125.

[IVAN76] Ivanilov, Y.I. and Mukhamediev, B.M. *An algorithm for solving the linear max-min problem.* Izv. Akad. Nauk SSSR, Tekhn. Kibernitika No. 6 (1976), 3-10.

[KALA82] Kalantari, B. and Rosen J.B. *Penalty for zero-one integer equivalent problems.* Math. Progr. 24 (1982), 229-232.

[KONN76] Konno, H. *Maximization of a convex quadratic function subject to linear constraints.* Math. Progr. 11 (1976), 117-127.

[LAWL63] Lawler, E.L. *The quadratic assignment problem.* Manag. Sc. 9 (1963), 586-599.

[MANG78] Mangasarian, O.L. *Characterization of linear complementarity problems as linear programs.* Math. Progr. Study 7 (1978), 74-88.

[PARD86] Pardalos, P.M. *Construction of test problems in quadratic bivalent programming.* Submitted to Discrete Appl. Math.

[RAGH69] Raghavachari, M. *On connections between zero-one integer programming and concave programming under linear constraints.* Oper. Res. 17 (1969), 680-684.

[THIE80] Thieu, T.V. *Relationship between bilinear programming and concave minimization under linear constraints.* Acta Math. Vietnam 5 (1980), 106-113.

Chapter 4 Enumerative methods in nonconvex programming

As we mentioned earlier an important property of concave functions is that every local and global solution is achieved at some extreme point of the the feasible domain. This property makes the problem more tractable since the search for global solutions can be restricted to the set of extreme points, even though this set in general may be too large to handle.

An obvious way to solve the concave programming problem, in the case where the feasible domain is a polyhedral set, is complete enumeration of the extreme points. Although most of the algorithms in the worst case will degenerate to complete inspection of all vertices of the polyhedron, this approach is computationally infeasible for large problems.

General techniques for total enumeration of the vertices of a linear polyhedron are given by [CHVA83] (chapter 18), Balinski [BALI61], Burdet [BURD74], Dyer and Proll [DYER77], Manas and Nedoma [MANA74], [DAHL75], [SCHM80], and Rossler [ROSS73]. The method of Manas and Nedoma is described in detail with numerical examples in the book by Martos [MART75, chapter 12]. A survey and comparison of methods for finding all vertices of a polyhedral set is given by Matheiss and Rubin [MATH80] and Dyer [DYER83]. From the computational complexity point of view, the problem of reporting the number of vertices of a polytope is #P-complete [LINI86], [VALI79].

Enumeration of the extreme points of a linear polyhedron is closely related to the problem of ranking the extreme points in ascending or descending order of the value of a linear function (or a linear fractional function [PARD86]) defined on the polyhedron.

4.1 Global concave minimization by ranking the extreme points

Consider the problem

$$\text{global} \min_{x \in P} f(x)$$

where the objective function $f(x)$ is concave and P is a bounded polyhedron in R^n.

Since the global minimum of this problem occurs at some vertex of P, one method for solving this problem is by searching among the extreme points of P using some linear underestimating function $g(x)$, that is $g(x) \leq f(x)$ for all $x \in P$. The first step in this approach is to rank the extreme points in increasing order of the value of $g(x)$ and then obtain lower and upper bounds on the global minimum in such a way that the lower bounds keep increasing and the upper bounds keep decreasing during the execution of the algorithm.

Cabot and Francis [CABO70] developed a procedure which solves concave quadratic problems using Murty's [MURT69] extreme point ranking algorithm applied to a certain linear underestimating function. The problem considered has the following form:

$$\min_{x \in S} f(x) = c^T x + x^T D x \tag{CP}$$

where $S = \{x: Ax = b, x \geq 0\}$ is a bounded linear polyhedron in R^n, A is an $m \times n$ matrix and $c \in R^n$, $b \in R^m$.

If $D = [d_1, \ldots, d_n]$ where d_i denotes the ith column of the matrix D, solve the following multiple-cost-row linear program

$$\min_{x \in S} d_i^T x, \; i=1, \ldots, n. \tag{MCR}$$

Let u_i be the minimum objective function value obtained for $i=1, \ldots, n$. Consider now the linear function

$$g(x) = \sum_{i=1}^{n} (c_i + u_i) x_i .$$

By construction and using the fact that $x^T D x = \sum_{i=1}^{n} (x^T d_i) x_i$, it is easy to check that

$$g(x) \leq f(x), \quad x \in S$$

that is, $g(x)$ is a linear underestimating function of $f(x)$ over the polyhedron P.

To obtain lower and upper bounds for the global minimum we solve the following linear program:

$$\min_{x \in S} g(x). \tag{LP}$$

Proposition 4.1.1: If x_0 is an optimal solution of (LP) and f^* is the global optimum of (CP), then

$$f_l = g(x_0) \le f^* \le f(x_0) = f_u.$$

To find these upper and lower bounds, we need to obtain the solution of $(n+1)$ linear programs. However all these linear programs have the same feasible domain, only the objective function changes. In special cases of problems, such as that of the quadratic assignment problem, finding the u_i's is often very simple (for the quadratic assignment problem only a simple rearrangement of the elements of two vectors is involved).

Since we consider only approximations of the quadratic part of the objective function, the obtained lower and upper bounds will often be good, if the linear term numerically dominates the nonlinear term. It remains an open question to obtain bounds on the approximation of the global optimum obtained by f_l and f_u. The following proposition is fundamental for the solution procedure.

Proposition 4.1.2: Given any upper bound f_u of f^*, denote by $\{x_k\}$ the set of all extreme points x_k of (LP) such that $g(x_k) \le f_u$. Then the original problem (CP) has a global solution x^* such that $x^* \in \{x_k\}$.

Proof: The global optimum of (CP) occurs at some vertex of S, and therefore is also a vertex of (LP). Since $g(x^*) \le f_u$, the conclusion now follows.

We describe next the procedure of [CABO70], using the linear underestimating function $g(x)$ and Murty's extreme point ranking approach:

1. Solve the linear program (LP) to obtain x_0; take $f_l = g(x_0)$ as a lower bound of the global minimum f^*.

2. Take $f_u = f(x_0)$ as an upper bound on f^*. Take x_0 as the "current best solution" to the original problem (CP).

3. Use Murty's extreme point ranking procedure to find the next best extreme point solution x_k to (LP). If $g(x_k) > f_u$, then stop; the current best solution is the

global solution to (CP), and $f^* = f_u$. If $g(x_k) \leq f_u$, then replace f_l by $g(x_k)$; f_l is a lower bound on f^*.

4. If $f(x_k) < f_u$, then replace f_u by $f(x_k)$ and replace the current best solution to (CP) by x_k; f_u is an upper bound on f^*. Otherwise return to step 3 without changing f_u or the current best solution.

A near optimum solution is obtained when the difference of the upper and lower bounds is small enough. The current best solution is a near optimum solution to the original problem.

In the above approach we considered Murty's extreme point ranking procedure. Murty's method is essentially based on the result that given the first, second , . . . , kth ranked extreme points of a linear programming problem, the next best $(k+1)$th ranked extreme point is geometrically adjacent to at least one of these k points (see also [MURT83]). Assuming that the problem is nondegenerate, the simplex algorithm may be used to maintain a list of such adjacent extreme points. In the presence of degeneracys however, one would also need to determine all basic representations of a degenerate vertex in order to use the simplex algorithm. Different methods have been proposed for handling the degenerate case and also a variety of methods have been developed for ranking the extreme points of polytopes with special structure (see for example [BAZA81]). Taha [TAHA73] also uses the idea of ranking the extreme points based on linear underestimating functions. The computational effectiveness of extreme point ranking algorithms is discussed in McKeown[McKE78].

4.2 Construction of linear underestimating functions

To use an extreme point ranking procedure we need methods of constructing linear underestimating functions. It is also obvious that a good approximation will reduce the computational effort considerably.

We consider here the problem of constructing linear underestimating functions to concave functions over bounded polytopes. Let $f(x)$ be a concave (not necessarily quadratic) function over a bounded polyhedron $P \subseteq R^n$, and let $V = \{v_1, \ldots, v_k, k \geq n+1\}$, be a set of points in R^n such that their convex hull, $Co(V) \supseteq P$. For example, in case $k = n+1$ the convex hull is a simplex, and if $k = 2^n$ it may be a rectangular domain.

We want to construct a linear function

$$\Gamma(x) = c^T x + \beta$$

where $c \in R^n$ and $\beta \in R$, such that $\Gamma(x)$ underestimates $f(x)$ on $Co(V)$ (and therefore on P), and such that $\Gamma(v_i) = f(v_i)$ for at least $n+1$ points of the set V. That is we want

$$\Gamma(v_i) \leq f(v_i), \quad i=1, \ldots, k \tag{LC}$$

with equality attained for at least $n+1$ of these. This can be accomplished by considering the following minimization problem:

$$\min_{c,\beta} \sum_{i=1}^{k} [f(v_i) - \Gamma(v_i)]$$

$$\text{s.t. } \Gamma(v_i) \leq f(v_i),\ i=1, \ldots, k$$

This problem is equivalent to

$$\max_{\beta, c} \bar{v}^T c + \beta \tag{LP}$$

$$\text{s.t. } \beta e_k + V^T c \leq b$$

where $\bar{v} = \frac{1}{k} \sum_{i=1}^{k} v_i$ is the centroid of the points v_i, $V = [v_1, \ldots, v_k]$, $b^T = (f(v_1), \ldots, f(v_k))$, and $e_k{}^T = (1, \ldots, 1) \in R^k$.

Considering this as the unsymmetric dual problem, we have the corresponding primal

$$\min_{z} b^T z \tag{PLP}$$

$$\text{s.t. } \sum_{i=1}^{k} z_i = 1,\ Vz = \bar{v},\ z \geq 0$$

Note that this linear programming problem has $n+1$ constraints (rows) and k variables. Its optimal basis will therefore contain $n+1$ columns, which will correspond to the active inequality constraints in the dual. Furthermore, each point v_i corresponds to a column in the primal constraint matrix, so that adding or deleting points is easily carried out.

Also note that a feasible (but non-basic) solution to (PLP) is given by $z = \frac{1}{k} e_k$, which gives $b^T z = \frac{1}{k} \sum_{i=1}^{k} f(v_i)$, the average of the function values. A feasible solution to the dual (LP) is given by $c = 0$, $\beta = \min_i f(v_i)$. Thus (PLP) has a bounded optimal solution. Now the optimal dual vector $w \in R^{n+1}$ for (PLP) gives β and c, that is $w^T = (\beta, c)$.

Other different methods for constructing linear underestimating functions are discussed later on. For the case of a quadratic concave objective function, we can use separable programming techniques for constructing "best" linear underestimating functions.

4.3 An algorithm for the indefinite quadratic problem

An algorithm proposed by Manas [MANA68] solves the general indefinite quadratic global optimization problem using enumeration of the vertices of the feasible domain. The disadvantage of this method is that can be used only on problems with a small number of variables and vertices. To keep the notation simple consider the problem

$$\text{global} \min_{x \in P} f(x) = x^T C x \qquad \text{(IQP)}$$

where $P=\{x \in R^n : Ax = b,\, x \geq 0\}$ is a bounded polyhedron, C is an $n \times n$ symmetric matrix, A is an $m \times n$ matrix and $b \in R^m$.

Theorem 4.3.1: The solution to (IQP) occurs on a boundary point of P, not necessarily a vertex.

Proof: Let x^* be a solution of (IQP), and suppose x^* is an interior point of P. Then $\nabla f(x^*) = 0$ and therefore $f(x^*) - f(x) = -(x-x^*)^T C (x-x^*)$ for all x. Since $f(x)$ is indefinite, there exists an $x_1 \in P$ such that $(x_1-x^*)^T C (x_1-x^*) < 0$, that is $f(x^*) > f(x_1)$ contradicting the assumption that x^* is a solution to (IQP).

It is easy to construct examples of (IQP) problems with nonextreme point solutions. However vertex enumeration plays an important role here.

The proposed algorithm consists of the following steps:

1. First find all vertices $v_1, v_2, \ldots, v_r$ of the polyhedron P, and then form the matrix

$$E = [v_1, v_2, \ldots, v_r]$$

2. Compute the matrix $D = E^T C E$. Let

$$Y = \{y \in R^r : \sum_{i=1}^{r} y_i = 1, \ y \geq 0\}$$

be a unit simplex in R^r. Then the following is true:

Proposition 4.3.1: Consider the problem

$$\min_{y \in Y} y^T D y = \bar{y}^T D \bar{y}.$$

Then

$$\min_{x \in P} x^T C x = \bar{x}^T C \bar{x}$$

where $\bar{x} = E\bar{y}$, and moreover $\bar{y}^T D \bar{y} = \bar{x}^T C \bar{x}$.

Proof: The proof follows easily from the fact that each point x in P is a convex combination of its extreme points.

Hence, instead of solving the original problem (IQP), it is sufficient to seek the minimum of the problem

$$\min_{y \in Y} y^T D y \qquad \text{(AP)}$$

$$\text{s.t.} \sum_{i=1}^{r} y_i = 1, \; y \geq 0$$

3. Choose an integer $p \geq 1$ and form the set

$$Z = \{z : z_i = a_i/p, \; i=1, \ldots, r, \; \sum_{i=1}^{r} a_i = p, \; a_i \in \{0, 1, \ldots, p\}\}$$

The set $Z \subseteq Y$ is a finite set with $\binom{p+r-1}{r-1} = \binom{p+r-1}{p}$ elements.

Next generate the points of Z in some order and for any $z \in Z$ compute the function value $z^T Dz$. Store only the point z_1 which has provided the smallest function value so far and also those points z_i, if any, for which $z_i^T Dz_i = z_1^T Dz_1$.

At the end of the computation we get a set of points z_i, $i=1, \ldots, s$ where $1 \leq s \leq \binom{p+r-1}{p}$, for which

$$z_i^T Dz_i = \min_{z \in Z} z^T Dz$$

The points z_i give approximate solutions to the (AP). Therefore, the points

$$x_i = Ez_i, \; i=1, \ldots, s$$

represent approximate solutions to the original (IQP) problem.

Although this approach is not computationally practical for large problems, a priori error bounds are easily obtainable.

Theorem 4.3.2: Let

$$\eta_j = \max_{1 \leq i \leq r} \{|d_{ij}|\}, \; j=1, \ldots, r$$

where d_{ij} are the elements of matrix D. Then for any $\bar{z}$ in the set $\{z_i, i=1,\ldots,s\}$ we have

$$0 \leq \bar{z}^T D\bar{z} - \bar{x}^T C\bar{x} \leq \frac{2}{p} \sum_{j=1}^{r} \eta_j$$

Proof: Let

$$\phi(\lambda) = (y+\lambda(z-y))^T D (y+\lambda(z-y)) = F(y+\lambda(z-y)),$$

where $z \in Z$ and $y \in Y$. By the mean value theorem, there exists a point $\xi \in (0, 1)$ such that

$$\phi(1)-\phi(0) = \phi'(\xi) = [(\frac{d}{d\lambda}F(y+\lambda(z-y))]_{\lambda=\xi}.$$

Putting $t = y+\lambda(z-y)$, we have

$$\phi(1)-\phi(0) = \sum_{j=1}^{r} [\partial F(t)/\partial t_j]_{\lambda=\xi} (\partial t_j/\partial\lambda),$$

that is

$$F(z)-F(y) = \sum_{j=1}^{r} [\partial F(t)/\partial t_j]_{\lambda=\xi} (z_j-y_j).$$

Since

$$\max_{t \in Y} |\partial F(t)/\partial t_j| = \max_{t \in Y} |2\sum_{i=1}^{r} d_{ij} t_i| \leq 2\max_{1 \leq i \leq r}\{|d_{ij}|\} = 2\eta_j$$

and

$$|F(z)-F(y)| \leq \sum_{j=1}^{r} (\max_{t \in Y} |\partial F(t)/\partial t_j|)|z_j-y_j|,$$

we have that

$$|F(z)-F(y)| \leq 2\sum_{j=1}^{r} \eta_j |z_j-y_j|,$$

for any points $z \in Z$, $y \in Y$.

Now let $\hat{z}$ denote the point of Z which is nearest to the point $\bar{y}$. Then $F(z) \leq F(\hat{z})$ and so

$$0 \leq F(\hat{z})-F(\bar{y}) \leq F(\bar{z})-F(\bar{y}) \leq \frac{2}{p}\sum_{j=1}^{r} \eta_j,$$

as $|\hat{z}_j-\bar{y}_j| \leq 1/p$, and this completes the proof.

Note that this error bound tends to zero as p increases. Also this is a worst case upper bound, and it is much larger than the usual actual error.

As we mentioned in an earlier chapter, an important class of indefinite quadratic problems is the Linear Complementarity Problem:

Given a matrix $M_{n \times n}$ and a vector $q \in R^n$,

find $x \geq 0$ such that $Mx+q \geq 0$, $x^T(Mx+q) = 0$

(or prove that such an x does not exist).

This problem can be formulated as an (indefinite) quadratic problem of the form:

$$\min_{x \in S} \phi(x) = x^T Mx + q^T x$$

$$\text{where } S = \{x : Mx+q \geq 0, x \geq 0\} \subseteq R^n$$

If this problem has a finite solution, then its solution is a vertex v of S, with $\phi(v) = 0$. Based on this fact several enumerative techniques have been proposed for the solution of linear complementarity problems. Two recent such algorithms are given in [AL-K85] and [JUDI85].

4.4 Concave cost network flow problems

A network is a directed graph $G(N,A)$, where $N=\{1, \ldots, n\}$ is the set of nodes and $A=\{(i,j) \in N \times N\}$ is the set of arcs. The set of nodes is partitioned into sources, sinks and intermediate nodes. Each arc is assigned a nonnegative number c_{ij}, called the capacity of the arc. The concave cost minimum flow problem can be formulated in the following way:

$$\min f(x) = \sum_{(i,j) \in A} f_{ij}(x_{ij})$$

$$\text{s.t. } 0 \leq x_{ij} \leq c_{ij} \quad (i,j) \in A$$

$$\sum_{j=1}^{n} (x_{ij} - x_{ji}) = b_i, \quad i=1, \ldots, n$$

where each $f_{ij}(x_{ij})$ is a concave function.

Problems of the above type have numerous applications, and are characterized by the special structure of the constraint set (network constraints), and their large size. In [FLOR71] a branch and bound algorithm is given, based on the equivalence of the general network flow problem to a network flow problem in a bipartite network of special form. An optimal flow is found by implicit enumeration of the set of extremal flows (vertices) in that network.

Zangwill [ZANG68], proposed an algorithm for the uncapacitated case that uses a characterization of the extreme points of the constraint set, and dynamic programming. More results on such characterizations of the extreme points and other properties that clarify the structure of these problems can be found in [LOZO82].

4.5 Exercises

1. Suppose that $v_1, \ldots, v_k$ are the first k vertices in a ranked sequence of extreme points. Prove that the next vertex v_{k+1} can be taken to be an adjacent vertex of $v_1, \ldots, v_k$, distinct from $v_1, \ldots, v_k$.

2. (i) Let $P = \{x: Ax \leq b\}$ be a bounded polyhedron in R^n, where A is an $m \times n$ matrix. Give an example of such a polyhedron with $O(2^m)$ vertices.
(ii) Let $Co(v_1, \ldots, v_k)$ be the convex hull of v_i, $1 \leq i \leq k$ in R^n. This is a bounded polyhedral set and can be defined as

$$Co(v_1, \ldots, v_k) = \{x: Ax \leq b\}, \ A \in R^{m \times n}.$$

Give an example where $m = \Omega(2^k)$ (and this is a minimal such possible representation).
Hint: Consider the $2n$ vectors of the form: $v_i =$ the ith column of the $n \times n$ identity matrix, $v_{n+i} = -v_i$, $i=1, \ldots, n$, and the 2^n constraints $\sum_{i=1}^{n} \varepsilon_i x_i \leq 1$, $\varepsilon_i = \pm 1$.

3. Consider the following global concave minimization problem

$$\min_x x^T Q x$$

$$\text{s.t. } x \in P = \{x : Ax \leq b, \ x \geq 0\} \subseteq R^n$$

where

$$A_{n \times n} = \begin{bmatrix} 1 & 2 & . & . & . & n-1 & n \\ 2 & 3 & . & . & . & n & 1 \\ . & . & . & . & . & . & . \\ . & . & . & . & . & . & . \\ . & . & . & . & . & . & . \\ n & 1 & . & . & . & n-2 & n-1 \end{bmatrix}, \quad b^T = (n(n+1)/2, \ldots, n(n+1)/2),$$

and Q is a tridiagonal matrix with diagonal elements $q_{ii} = -2$ and the next to diagonal entries equal to 1.

a) Find all vertices of the polyhedron P.

b) Prove that this problem has n local (global) minima with the same objective function value. Find these optima.

4. Given a matrix $M_{n \times n}$ and a vector $q \in R^n$ consider the polytope $S = \{x : Mx + q \geq 0, x \geq 0\}$.

a) Prove that S has at most 2^n vertices.

b) If S is nonempty and M is a negative definite matrix then prove that S is bounded.

c) For what matrices M is the polytope S nonempty and bounded?

5. The linear fractional programming problem can be stated as follows:

$$\min_{x \in P} f(x) = \frac{c^T x + \alpha}{d^T x + \beta}$$

where $P = \{x : Ax \leq b, x \geq 0\}$ is a bounded polyhedron in R^n, A is an $m \times n$ matrix, $c, d \in R^n$, $b \in R^m$ and α, β are scalars. Assume that $d^T x + \beta \neq 0$ for all $x \in P$.

a) Prove that $f(x)$ has a minimum at an extreme point of P.

b) Let the set of vertices $x_1, x_2, \ldots, x_k$ be ranked such that $f(x_1) \leq f(x_2) \leq \cdots \leq f(x_k)$. Then the next vertex in the sequence, x_{k+1}, must be adjacent to one of $x_1, \ldots, x_k$.

c) Using the above result, modify a vertex ranking algorithm for linear programming to rank the vertices in the linear fractional programming problem

([STOR83], [PARD86]).

4.6 References

[AL-K85] Al-Khayyal, F.A. *An implicit enumeration algorithm for the general linear complementarity problem.* Tech. Report, School of Indust. and Syst. Engin., The Georgia Inst. of Technology (1985).

[BALI61] Balinski, M.L. *An algorithm for finding all vertices of polyhedral sets.* J. Soc. Indust. Appl. Math. 9 (1961), 72-89.

[BAZA81] Bazara, M.S. and Sherali, H.D. *A versatile scheme for ranking the extreme points of an assignment polytope.* Naval Research Log. Quart. Vol. 28 (1981), No. 4.

[CABO70] Cabot, V.A. and Francis, R.L. *Solving certain nonconvex quadratic minimization problems by ranking the extreme points.* Operations Research, Vol. 18 (1970), 82-86.

[CHVA83] Chvatal, V. *Linear Programming.* Freeman N.Y. (1983)

[BURD74] Burdet., C.A. *Generating all faces of a polyhedron.* SIAM J. Appl. Math. 26 (1974), 72-89.

[DAHL75] Dahl, G. and Storoy, S. *Decomposed enumeration of extreme points in the linear programming problem.* BIT 15 (1975), 151-157.

[DYER83] Dyer, M.E. *The complexity of vertex enumeration methods.* Math. Oper. Res. 8 (1983), 381-402.

[DYER77] Dyer, M.E. and Proll, L.G. *An algorithm for determining all extreme points of a convex polytope.* Math. Progr. 12 (1977), 81-96.

[FLOR71] Florian, M. and Robillard, P. *An implicit enumeration algorithm for the concave cost network flow problem.* Manag. Science, Vol 18 (1971), 184-193.

[JUDI85] Judice, J.J. and Mitra, G. *An enumerative method for the solution set of linear complementarity problems.* Working Paper, Math. Dept., Brunel Univer. (1985).

[LINI86] Linial, N. *Hard enumeration problems in geometry and combinatorics.* SIAM J. Alg. Discr. Meth. Vol. 7, No 2 (1986), 331-335.

[LOZO82] Lozovanu, D.D. *Properties of optimal solutions of a grid transport problem with concave cost function of the flow on the arcs.* Engin. Cybernetics Vol. 20 (1982), 34-38.

[MATH80] Matheiss, T.H. and Rubin, D.S. *A survey and comparison of methods for finding all vertices of convex polyhedral sets.* Math. Oper. Reser. 5 (1980), 167-185.

[McKE78] McKeown, P.G. *Extreme point ranking algorithms: A computational survey.* In Computers and Math. Progr., W.W. White ed., National Bureau of Standards Special Publication 502 (1978), U.S. Government Printing Office, Washington DC, 216-222.

[MANA68] Manas, M. *An algorithm for a nonconvex programming problem.* Econom. Math. Obzor, Acad. Nacl. Ceskoslov. (1968), 202-212.

[MANA74] Manas, M. and Nedoma, J. *Finding all vertices of a convex polyhedral set.* Numer. Mathem. 9 (1974), 35-39.

[MART75] Martos, B. *Nonlinear Programming: theory and methods* North-Holland Publishing Company, (1975).

[MURT68] Murty, K.G. *Solving the fixed charge problem by ranking the extreme points.* Operations Research, Vol 18 (1968), 268-279.

[MURT83] Murty, K.G. *Linear Programming* Wiley (1983).

[PARD86] Pardalos, P.M. *An algorithm for a class of nonlinear fractional problems using ranking of the vertices.* BIT 26 (1986), 392-395.

[ROSS73] Rossler, M. *A method to calculate an optimal production plan with a concave objective function.* Unternehmnsforschung, 20 (1973), 373-382.

[SCHM80] Schmidt, B.K. and Mattheiss, T.H. *Computational results on an algorithm for finding all vertices of a polytope.* Math. Progr. Vol. 18 (1980), 308-329.

[STOR83] Storoy, S. *Ranking of vertices in the linear fractional programming problem.* BIT 23 (1983), No. 3, 403-405.

[TAHA73] Taha, H.A. *Concave minimum over a convex polyhedron.* Naval Res. Logist. Quart. Vol. 20 (1973), 353-355.

[VALI79] Valiant, L.J. *The complexity of enumeration and reliability problems.* SIAM J. Comput., Vol 8 (1979), 410-421.

[ZANG68] Zangwill, W.I. *Minimum concave cost flows in certain networks.* Manag. Science. Vol 14 (1968), 429-450.

Chapter 5 Cutting plane methods

Cutting plane methods have been used in optimization for some time. A cutting plane (or simply a cut) is a linear constraint that is used to reduce the constraint set in such a way that does not exclude optimal feasible points. However cutting planes are also used to obtain bounds for the global optimum in conjunction with other techniques. The main difference between the various cutting plane approaches is the generation of cuts and the way the feasible domain is updated when a new cut is determined.

5.1 Tuy's "cone splitting" method

H. Tuy's original method [TUY64] is concerned with finding the global optimum of the following problem:

$$\text{global} \min_{x \in P} f(x)$$

where $f(x)$ is a concave function and P a polyhedron in R^n.

Before stating the algorithm we present some of the main ideas. In this approach the constraint set P can be thought as contained in a cone generated by the edges coincident with a vertex. The method solves a sequence of subproblems associated with subcones of this initial cone. Cutting plane techniques are used to obtain bound estimates of the global optimum.

Given a set of linearly independent vectors $u_1, \ldots, u_n$ in R^n the cone generated by these vectors is the set of all convex combinations of $u_1, \ldots, u_n$ with nonnegative coefficients, that is,

$$C(u_1, \ldots, u_n) = \{x: x = \sum_{i=1}^{n} \lambda_i u_i,\ \lambda_i \geq 0\}$$

If $U = [u_1, \ldots, u_n]$ then it is easy to see that we can write

$$C = \{x: U^{-1}x \geq 0\}$$

Also the hyperplane passing through $u_1, \ldots, u_n$ is given by

$$H = \{x: e^T U^{-1} x = 1\}$$

where $e^T = (1, \ldots, 1) \in R^n$.

Let $S(u_1, \ldots, u_n)$ denote the simplex that is formed when the cone C is cut off by the hyperplane H. Consider now the auxiliary linear program

$$A(u_1, \ldots, u_n): \quad \max_{x \in P} e^T U^{-1} x$$

What is the relation between this auxiliary linear program and the original problem? To understand this relation assume that $0 \in P$ is a vertex and that we consider the cone generated by 0 and the u_i's, $i=1, \ldots, n$, vectors on the coordinate axes (i.e. the coordinate axes are edges of P emanating from 0). Note that $P \cap C \subseteq P \cap S$ if the solution of the auxiliary problem has objective function value ≤ 1. In that case, since $f(x)$ is concave,

$$f(x) \geq \min(f(u_1), \ldots, f(u_n), f(0)) \quad \text{for all } x \in P \cap C$$

Tuy's method involves covering P with such cones and solving subproblems associated with these cones.

We describe next this method ([TUY64], [ZWAR73]):

1. Find an initial vertex $x_0 \in P$ such that $f(x) \geq f(x_0) = \alpha_0$ for any neighboring vertex x. Without loss of generality we may assume that x_0 is the origin. We may also assume that x_0 has n neighboring vertices (nondegeneracy is assumed), each of which is proportional to a unit vector in R^n (i.e. the neighboring vertices are on the n coordinate axes).

2. Set up the first auxiliary linear problem. Let u_k be the direction vector along the kth edge issuing from x_0. Let

$$\theta_{1,k} = \max\{\theta: f(\theta u_k) \geq \alpha_0\}$$

In case of unboundness choose $\theta_{1,k}$ as large as convenient. Let

$$y_{1,k} = \theta_{1,k} u_k, \quad k=1, \ldots, n$$

Then the first auxiliary linear problem is $A_{1,1}(y_{1,1}, \ldots, y_{1,n})$. Set $q = 1$.

3. Given α_{q-1} and the auxiliary problems $A_{q,1}, \ldots, A_{q,k_q}$, start step q as follows:

a) Solve all auxiliary linear problems.

b) Set $\alpha_q = \min\{\alpha_{q-1}, f(v)\}$ for all vertices v, encountered while solving the above auxiliary problems.

c) For each $A_{q,k}$ that has objective function value >1, generate new auxiliary problems in the following way:

If $x_{q,k}$ is the solution of $A_{q,k} = A(y_{q,k_1},\ldots,y_{q,k_n})$ and $x_{q,k} = \sum_{i=1}^{n} \lambda_i^{q,k} y_{q,k_i}$ (where $\lambda_i^{q,k} \geq 0$ and $\sum_{i=1}^{n} \lambda_i^{q,k} = 1$) then set

$$\overline{x}_{q,k} = \theta_{q,k} x_{q,k}, \ \overline{y}_{q,k_i} = \theta_{q,k_i} y_{q,k_i}, \ i=1, \ldots, n$$

where

$$\theta_{q,k} = \max \{\theta: f(\theta x_{q,k}) \geq \alpha_q\} \text{ and } \theta_{q,k_i} = \max \{\theta: f(\theta y_{q,k_i}) \geq \alpha_q\}$$

Then, for each nonzero $\lambda_i^{q,k}$, create the auxiliary problem

$$A(\overline{y}_{q,k_1}, \ldots, \overline{y}_{q,k_{i-1}}, \overline{x}_{q,k}, \overline{y}_{q,k_{i+1}}, \ldots, \overline{y}_{q,k_n})$$

The collection of these new auxiliary problems is the q+1st set. If this set is empty, stop; α_q is the global minimum. Otherwise go to 3.

Remarks: Geometrically, we can say that the cone $C(y_{q,k_1}, \ldots, y_{q,k_n})$ is replaced by a set of cones $C(\overline{y}_{q,k_1}, \ldots, \overline{x}_{q,k}, \ldots, \overline{y}_{q,k_n})$, whose union contains $C(y_{q,k_1}, \ldots, y_{q,k_n})$ (cone splitting). The new auxiliary problems associated with these new cones have more chance of yielding an optimal objective value <1, and therefore of being excluded from further consideration.

It is important to notice that all auxiliary problems in Tuy's method have the same constraint set, and in particular the constraints of the original problem. The linear functions are easily calculated by means of the inverse U^{-1}. If an auxiliary problem under consideration gives rise to new ones, the new matrix $\overline{U}$ is obtained from the old simply by multiplying each column by some number and replacing one column. Hence knowing U^{-1}, we can easily find $\overline{U}^{-1}$ by well known formulas, for example the multiplicative form of the inverse matrix.

There is also the question of finding the first vertex x_0. This vertex can be found by using the Frank-Wolfe algorithm [FRAN56]. If x_0 is a vertex we solve the linear program

$$\min_{x \in P} \nabla f(x_0)^T x.$$

If x_0 solves this linear program then check if it satisfies $f(x_0) \leq f(x)$ for all adjacent vertices x. If for some neighbor vertex z we have $f(z) < f(x_0)$ then set $x_0 \leftarrow z$ and return to the linear program. On the other hand if x_0 is not a solution, let x_0 be any solution and return to the linear program. Note that at this point nondegeneracy is important. It is also possible that an exponential number of steps may be needed to find an initial suitable vertex x_0.

The cone splitting algorithm described above may not converge, as has been shown by [ZWAR73]. Zwart gives an example in which the sequence of subproblems is nonterminating because of cycling. Several authors ([ZWAR74], [BALI78], [KRYN79], [BULA82] [INDI84]) developed methods by modifying Tuy's approach to obtain a convergent procedure or to get approximate solutions. Also [JACO81] gave a proof of a Tuy-type algorithm for the general concave problem.

5.2 General class of algorithms

A class of algorithms which are based upon a combination of branch and bound techniques with Tuy's original method are proposed by Thoai and Tuy [THOA80].

Assume that $\{x: f(x) \geq \alpha\}$ is bounded for any $x \in R^n$ and a real number α, and the polyhedron is of the form $P = \{x: Ax \leq b\}$. In addition we assume that the origin $v_0 = 0$ is a nondegenerate vertex of P. If $v_1, \ldots, v_n$ are its adjacent extreme points (which are linearly independent) denote by C_0 the cone generated by n rays emanating from $v_0 = 0$ and passing through $v_1, \ldots, v_n$ respectively. The general method proposed by [THOA80] can be described as follows.

Starting from C_0 compute a suitable lower bound $\mu(C_0)$ for $f(x)$ over $P \cap C_0$. Let x_0 be the vertex defined by

$$f(x_0) = \min_{0 \le i \le n} \{f(v_i)\}$$

Let $\alpha_0 = f(x_0)$ and $F_0 = \{C_0\}$.

At step $k=0, 1, \cdots$ we have a best current solution x_k such that $f(x_k) = \alpha_k$, and a set of cones F_k, such that every member of F_k is either a cone in F_{k-1} or a subcone of a member in F_{k-1} ($k >= 1$). With each cone C in F_k is associated a number $\mu(C)$ which is a lower bound for $f(x)$ over $P \cap C$.

Let $\overline{F}_k = \{C \in F_k : \mu(C) < \alpha_k\}$ (delete all cones C from F_k for which $\mu(C) \ge \alpha_k$). If $\overline{F}_k = \emptyset$ stop; x_k is the global minimum. Otherwise let $C \in \overline{F}_k$ with smallest $\mu(C)$. Split this cone into smaller cones $C_{k,1}, \ldots, C_{k,r_k}$ and for each $i=1, \ldots, r_k$ compute $\mu(C_{k,i})$, lower bounds of $f(x)$ over $P \cap C_{k,i}$. Let x_{k+1} be the best current solution with $f(x_{k+1}) = \alpha_k$. Form the new set F_{k+1} by replacing C_k with its subcones $C_{k,1}, \ldots, C_{k,r_k}$ and then go to step $k+1$.

In this general approach a cone containing the feasible domain P is successively subdived into smaller subcones and then bound estimates over these subcones are obtained using cutting plane techniques developed by [TUY64]. In order to ensure convergence, a branching process is built in such a way that when we do not have termination in a finite number of steps, a sequence of subcones is generated that tends to a ray. More recently [TUY85], a method was proposed based on a similar cone splitting scheme with a different bounding procedure which does not require the solution of auxiliary problems at each step. The main advantage of this new bounding procedure is that it can be applied to convex as well as to linear constraints. However there is limited computational experience with these methods, and preliminary numerical experiments indicate that only moderate size problems can be solved.

The Tuy cut was generalized by Glover[GLOV73A] to give a broader class of cuts called convexity cuts. In addition Glover describes a new approach to generate cuts (cut search). Some discussion on parallel developments in concave and integer programming can be found in [GLOV73B].

In the above approaches using Tuy's cone splitting with cut generation method, we solve a sequence of subproblems to find the global optimum. In the process of solution we have a sequential replacement of problems by subproblems. Other approaches have been proposed where we have a sequential reduction of the feasible domain by a sequence of cutting planes. One of these methods was proposed by Ritter ([RITT65], [RITT66]) for the general indefinite quadratic problem under linear constraints. Zwart [ZWAR73] gave a counterexample in which the sequence of cutting planes becomes interminable before the global minimum is found.

5.3 Outer and inner approximation methods

When the feasible region is a general convex set in R^n the problem becomes computationally very difficult. In a paper by Tuy [TUY83], a method is discussed that involves ideas that underlie the cutting plane method of Kelly [KELL60] for solving convex programs. Basically the method consists of approximating the given feasible domain by a linear polyhedron containing it (outer approximation), in such a way that the vertices of the polyhedron are known or can be computed practically. Then the minimum of the objective function over this polyhedral set gives an approximate solution to the original problem. If the solution attained is not yet satisfactory, the approximation is refined further, and the whole process of successive approximations can be arranged so that it will converge to an optimal solution of the original problem. Such approximation procedures have been combined with objective function underestimators that we are going to discuss later on.

A different approximate procedure is considered by Mukhamediev [MUKH82]. Here a new approximate algorithm is proposed which permits one to obtain an approximate optimal solution with a given accuracy. The idea of the proposed method is to approximate from the interior (inner approximation) the feasible domain by polytopes. Under some additional assumptions on the objective function, it is proved that after a finite number of steps the algorithm generates an ε-approximate solution to the global optimum.

5.4 Exercises

1. Let $u_1, \ldots, u_n$ be linearly independent vectors in R^n. Let C be the cone generated by these vectors and let z be a nonzero vector in C. Define $P(z) = \{i : \lambda_i > 0\}$ where $\lambda \geq 0$ is the vector of weights satisfying $z = \sum_{i=1}^{n} \lambda_i u_i$. Let C_i be the cone generated by $u_1, \ldots, u_{i-1}, z, u_{i+1}, \ldots, u_n$. Then prove that $C = \bigcup_{i \in P(z)} C_i$.

2. Consider the problem

$$\text{global} \min_x f(x) = -x^2 - y^2 - (z-1)^2$$

subject to

$$-y \leq 0, \quad x+y-z \leq 0, \quad -x+y-z \leq 0, \quad -6x+y+z \leq 1.9$$

$$12x+5y+12z \leq 22.8, \quad 12x+12y+7z \leq 17.1$$

Apply Tuy's algorithm to this problem using the origin as the initial vertex x_0 [ZWAR73].

3. Consider the assignment polytope

$$P = \{x : \sum_{i=1}^{n} x_{ij} = 1, \; j=1, \ldots, n, \; \sum_{j=1}^{n} x_{ij} = 1, \; i=1, \ldots, n, \; 0 \leq x_{ij} \leq 1\}$$

i). Let v be any extreme point of P. How many extreme points adjacent to v are there?

ii). Suppose we consider the problem of finding the global minimum of a concave separable function over P. Discuss a method for obtaining an initial vertex x_0 when Tuy's method is applied.

5.5 References

[BALI78] Bali, S. and Jacobsen, S. *On the convergence of the modified Tuy algorithm for minimizing a concave function on a bounded convex polyhedron.* Proceedings of the 8th IFIP Conf. on Optim. Techniques, Würzburg, Optimization Techniques (ed. J. Stoer), Springer-Verlag (1978), 59-66.

[BULA82] Bulatov, V.P. and Kasinskaya, L.I. *Some methods of concave minimization on a convex polyhedron and their applications* (Russian). Methods of optimization and their applications Nauka Sibirsk, Otdel. Novosibirsk. (1982), 71-80.

[FRAN56] Frank, M. and Wolfe, P. *An algorithm for quadratic programming. Naval Res. Log.* Quarterly 3 (1956), 95-110.

[GLOV73A] Glover, F. *Convexity cuts and cuts search.* Oper. Res. 21 (1973), 123-134.

[GLOV73B] Glover, F. and Klingman, D. *Concave programming applied to a special class of 0-1 integer programs.* Oper. Res. 21 (1973), 135-140.

[INDI84] Indihar, S. *Maximization of a convex function on a bounded polytope.* IV Conference on Applied Math. (Split 1984), 131-138.

[JACO81] Jacobsen, S.E. *Convergence of a Tuy-type algorithm for concave minimization subject to linear inequality constraints.* Appl. Math. Opti. 7 (1981), 1-9.

[KELL60] Kelly, J.E. *The cutting plane method for solving convex programs.* J. Soc. Industr. Appl. Math. 8 (1960), 703-712.

[KRYN79] Krynski, S.L. *Minimization of a concave function under linear constraints (modification of Tuy's method).* Survey of Mathematical Programming (Proc. Ninth International Progr. Sympos. Budapest 1976) Vol. 1,North Holland, Amsterdam (1979), 479-493.

[MUKH82] Mukhamediev, B.M. *Approximate methods for solving concave programming problems.* USSR Comput. Maths. Math. Phys. 22 (1982), 238-245.

[RITT65] Ritter, K. *Stationary points of quadratic maximum problems*. Z. Wahrscheinlichkeitstheorie verw. Geb. 4 (1965), 149-158.

[RITT66] Ritter, K. *A method for solving maximum problems with a nonconcave quadratic objective function*. Z. Wahrscheinlichkeitstheorie verw. Geb. 4 (1966), 340-351.

[THOA80] Thoai, Ng.V. and Tuy, H. *Convergent algorithms for minimizing a concave function*. Math. of Oper. Research Vol.5 (1980), 556-566.

[TUY64] Tuy, H. *Concave programming under linear constraints*. Dokl. Akad. Nauk SSR (1964), 159, 32-35.

[TUY83] Tuy, H. *On outer approximation methods for solving concave minimization problems*. Report No. 108 (1983), Forschungsschwerpunkt Dynamisch Systeme Universität Bremen, West Germany.

[TUY85] Tuy, H., Thieu, T.V., and Thai, Ng.Q. *A conical algorithm for globally minimizing a concave function over a closed convex set*. Math. of Oper. Research, Vol. 10, No. 3 (1985), 498-514.

[ZWAR73] Zwart, P.B. *Nonlinear Programming: Counterexamples to two global optimization algorithms*. Oper. Reser. 21 (1973), 1260-1266.

[ZWAR74] Zwart, P.B. *Global maximization of a convex function with linear inequality constraints*. Oper. Reser. 22 (1974), 602-609.

Chapter 6 Branch and Bound methods

Branch and bound techniques are the most commonly used for the efficient solution of nonconvex global optimization problems. Branching usually refers to a successive partioning of the feasible domain, and bounding refers to the determination of lower and upper bounds for the global optimum. Branch and bound techniques differ in the way they define rules for partitioning and the methods used for deriving bounds, avoiding exhaustive search.

6.1 Rectangular partitions and convex envelopes

Falk and Soland [FALK69] considered the global minimization of the following class of nonconvex separable problems:

$$\min f(x) = \sum_{i=1}^{n} f_i(x_i)$$

$$\text{s.t. } x \in D \cap C \subseteq R^n$$

where D is a compact convex set and $C = \{x : l_i \leq x_i \leq L_i,\ i=1, \ldots, n\}$ is a rectangular domain.

Separable programming contains a large class of problems including quadratic programming. Problems with a quadratic objective function can be reduced to an equivalent separable program using a linear transformation. One such reduction is going to be examined later on.

Crucial to the algorithm proposed by Falk and Soland is the use of convex envelopes. Suppose that ϕ_i is the convex envelope of f_i on the interval $[l_i, L_i]$; then $\phi = \sum_{i=1}^{n} \phi_i$ is the convex envelope of $f(x)$ on the rectangle C. The algorithm proposed here is of the branch and bound type, and solves a sequence of subproblems in each of which the objective function is linear or convex. These problems correspond to successive partitions of the feasible domain. Two different rules lead to convergence of the algorithm under certain requirements on the problem functions.

A brief description of the main ideas of the algorithm is given next. The present exposition was influenced by that of [McCO72]. Details and some illustrative examples are given in [FALK69]. Although the algorithm can be applied to a large class of objective functions (at least from the theoretical point of view) we assume here that $f(x)$ is a concave separable function and the feasible domain is a polyhedron.

The algorithm generates a sequence of points x_k, each of which is a solution of a problem P_{kv} defined as follows: Let

$$C_{kv}=\{x: l^{kv} \leq x \leq L^{kv}\}$$

In problem P_{kv}, the function f_j is replaced by its convex envelope ϕ_j^{kv} over the interval $[l_j^{kv}, L_j^{kv}]$. Then

$$\phi^{kv}(x) = \sum_{j=1}^{n} \phi_j^{kv}(x)$$

is the convex envelope of f over C_{kv}. Note that the convex envelope of a separable concave function is a linear function. The subproblem P_{kv} associated with any rectangle C_{kv} is then

$$\min \phi^{kv}(x)$$

$$\text{s.t. } x \in D \cap C_{kv}.$$

If x^{kv} is a solution point to P_{kv} then $\phi^{kv}(x^{kv})$ is a lower bound to f over $D \cap C_{kv}$.

At step k of the algorithm, the original rectangle C has been subdivided into p_k rectangles $\{C_{k1}, \ldots, C_{kp_k}\}$ that form a partition of C. Associated with each rectangle C_{kv} is a convex underestimating function ϕ^{kv} and a convex (linear when f concave) programming problem P_{kv} with solution point x^{kv}. Consider the rectangle with smallest objective function value. That is, let v_k denote the index such that

$$\phi^{kv_k}(x^{kv_k}) = \min_v \phi^{kv}(x^{kv}), \quad v=1, \ldots, p_k.$$

To make the notation simple, let x^k denote x^{kv_k}. The termination rule for the algorithm is that if $f(x^k) = \phi^{kv_k}(x^k)$, then the problem is solved by x^k. This is true because

$$\phi^{kv_k} \le \phi^{kv}(x) \le f(x)$$

for all $x \in D \cap C_{kv}$ and $v=1, \ldots, p_k$. If $f(x^k) > \phi^{kv_k}(x^k)$ the algorithm proceeds to step $k+1$ by dividing the rectangle C_{kv_k} into two or more rectangular subsets.

Next we state the branching rules for the general problem. The form of the rules depends on whether or not the problem functions are continuous or lower semicontinuous.

Weak branching rule: Choose any j that minimizes the difference

$$f_j(x_j^k) - \phi_j^{kv_k}(x_j^k)$$

Then $p_{k+1} = p_k + 1$ i.e. two new rectangles are formed by splitting C_{kv_k} into two parts. The bounds for both new rectangles are the same for all components except the jth which in one case has $[l_j^{kv_k}, x_j^k]$ as its bounds, and in the other, $[x_j^k, L_j^{kv_k}]$.

Strong branching rule: For every j such that

$$f_j(x_j^k) - \phi_j^{kv_k}(x_j^k) > 0$$

divide the corresponding interval $[l_j^{kv_k}, L_j^{kv_k}]$ into two intervals $[l_j^{kv_k}, x_j^k]$ and $[x_j^k, L_j^{kv_k}]$, creating a new rectangle for step $k+1$ for every j.

Note that the strong branching rule in general generates many more new rectangles, and hence more subproblems than the weak branching rule. However, it may be needed to guarantee convergence of the algorithm. Two statements about convergence are stated in the following theorems [FALK69].

Theorem 6.1.1: If the strong branching rule is used to generate new rectangles, then any limit point of x^k is a global minimum.

Theorem 6.1.2: If the function f is continuous and if the weak branching rule is used to generate new rectangles, then every limit point of x^k is a global minimum.

The algorithm described above has been extended later by Soland [SOLA71] to handle separable nonconvex constraints. An approximate algorithm for the separable nonconvex problem was proposed by Falk [FALK72], and coded by Grotte [GROT76]. This algorithm provides approximate solutions by replacing each of the original functions with their piecewise linear convex envelope. A branch and bound procedure first solves this approximate problem to get estimates on the optimal value of the approximating function, and then to set up new problems if the estimates do not yield a global solution.

In general the construction of the convex envelope of a continuous function is a very difficult computational problem, and the refining rules given by Falk and Soland's algorithm are such that none of the convergence theorems hold if one is using other functions instead of convex envelopes. Horst [HORS78], proposed a different approach for the separable nonconvex problem. Instead of considering convex envelopes and minimizing several convex subproblems, he only minimizes each $f_i(x_i)$ over suitable subintervals. He gives also a heuristic method for obtaining the global minimum of a function of one variable.

6.2 Simplex partitions

A more general algorithm of the branch and bound type is proposed by [HORS76] for the following general nonconvex programming problem:

$$\min \{f(x): x \in D \subseteq R^n\}$$

where D is a compact set and $f(x)$ is a continuous function. The algorithm solves a sequence of subproblems in each of which the objective function is convex or even linear. The difference between this approach and other approaches (as for example [FALK69]) is the use of general compact partitions instead of rectangular ones and a different refining rule such that the algorithm does not rely on the concept of convex envelopes and handles nonseparable functions. The algorithm generates a sequence of points x_k such that at least one accumulation point solves the problem. In a later paper [HORS80] it is shown that every accumulation point of the sequence x_k solves the problem.

Although the algorithm developed by Horst can be used for a general nonconvex function and compact partitions, we give here a description for the concave minimization problem where also the feasible domain is convex. The key idea of the algorithm is to use suitable simplices as partitions over which linear subproblems are defined. Initially a simplex must be found that contains D. The proposed algorithm is based on the following results:

Theorem 6.2.1: Let $S = [v_0, \ldots, v_n]$ be a simplex in R^n generated by the (affinely independent) vectors $v_0, \ldots, v_n$. If $v \in S$ ($v \neq v_i$) then

$$v = \sum_{i=0}^{n} \lambda_i v_i, \quad \sum_{i=0}^{n} \lambda_i = 1, \ \ \lambda_i \geq 0, i=0, \ldots, n$$

Replacing one vertex v_i of S with $\lambda_i > 0$ by v we obtain a subsimplex

$$S_i = [v_0, \ldots, v_{i-1}, v, v_{i+1}, \ldots, v_n] \subseteq S$$

The set of all subsimplices S_i (constructed by means of v) forms a partition of S.

Theorem 6.2.2: Let S_1 be a simplex in R^n. Construct a sequence of simplices S_k, $k=1, 2, \cdots$ in the following way: To construct S_{k+1} from S_k choose v to be the midpoint of the longest edge of S_k and let S_{k+1} be one of the two subsimplices constructed according to the above theorem. Then there is an $x \in R^n$ satisfying

$$\lim_{k \to \infty} S_k = \bigcap_{k=1}^{\infty} S_k = \{x\}$$

Proof: Exercise

As we proved earlier the convex envelope of a concave function $f(x)$ over a simplex S in R^n is a linear function $L(x) = c^T x + b$ which agrees with $f(x)$ at all vertices of S. Then the following result is true:

Theorem 6.2.3: Let S_1 be a simplex in R^n, and let S_k, $k=1, 2, \ldots$ be a sequence of simplices constructed from S_1 according to the previous theorem. Furthermore let $f: A \to R$ be a concave function on an open set A, $S_1 \subseteq A \subseteq R^n$, and let L_k, $k=1, 2, \cdots$ be the sequence of convex envelopes (linear functions) of f over

S_k. Then we have

1. $L_{k-1}(x) \le L_k(x) \le f(x)$, for all $x \in S_k$, k=1, 2, $\cdots$
2. For every ε>0 there exists a δ not depending on k such that

$$|L_k(x) - L_k(y)| \le \varepsilon$$

whenever $x, y \in S_k$ and $|x-y| \le \delta$ (this means that the sequence of functions is equicontinuous).

The steps of the proposed algorithm are suggested by the above theorems. Initially a simplex S_1 is constructed that contains the feasible domain D. A simple method to construct such a simplex is the following: First we consider the linear program

$$z_0 = \max_{x \in D} \sum_{i=1}^{n} x_i .$$

Then take S_1 to be the simplex $S = [v_0, \ldots, v_n]$, where $v_0 = 0$, the origin, and $v_i = z_0 e_i$, i=1, ..., n, where e_i denotes the ith unit-coordinate vector. Next we describe Horst's algorithm:

Step 0: Set $I^{(1)} = \{1\}$ and let $L_1(x)$ be the linear function (convex envelope) corresponding to the simplex S_1; go to step 1.

Step k: For each $i \in I^{(k)}$ solve the linear program

$$L_i(x_i) = \min_{x \in D \cap S_i} L_i(x)$$

Let $j \in I^{(k)}$ be the index for which $L_j(x_j) = \min_{i \in I^{(k)}} \{L_i(x_i)\}$.

If $f(x_j) = L_j(x_j)$, stop the algorithm; x_j solves the problem.

If $L_j(x_j) < f(x_j)$ then partition S_j into two subsimplices as described by Theorem 6.2.2 (call them S_l and S_k for l,k not in $I^{(k)}$). Let $L_l(x)$ and $L_k(x)$ be the corresponding linear functions. Set

$$I^{(k+1)} = (I^{(k)} \cup \{l, k\}) - \{j\}.$$

Go to step k+1.

Note that for each $i \in I^{(k)} \cap I^{(k-1)}$, the linear program was already solved at step $k-1$. Hence, there are at most two linear programs to be solved at each step.

If the solution has not been found after a finite number of steps, then it guarantees that every accumulation point of the sequence x_k is a solution. For more details see [HORS76].

Benson [BENS82A] presents a new convergence property for each of the two branch and bound algorithms for nonconvex programming problems developed by Falk-Soland and Horst. A prototype branch and bound algorithm is proposed which also uses underestimating subproblems. The Falk, Soland, and Horst algorithms are special cases of this prototype algorithm (see also [BENS82B] [BENS85]). More recently [HORS85], a more general class of branch and bound algorithms have been proposed. These algorithms are based on the previous approaches and give some general convergence results.

6.3 A successive underestimation method

Falk and Hoffman ([FALK76], [HOFF75]) proposed a finite and more efficient method for the concave minimization problem

$$\min_x f(x)$$

$$\text{s.t.} \quad x \in P = \{x: Ax \leq b\} \subseteq R^n$$

based on convex underestimation. Their method may be considered a variation of the Falk-Soland algorithm for separable nonconvex problems, although here separability is not required. In this method, the convex envelope of the objective function taken over a set containing the compact feasible domain is generated at each step, and minimized over the feasible domain. The convex envelope is then refined by tightening the containing set, and the process continues until the covering set is sufficiently close to the feasible domain. The algorithm is finite and is guaranteed to terminate at the global optimum. A description of the method is given next.

Initially the algorithm selects a (nondegenerate) vertex v_0 by the first phase of the simplex method. Then a simplex S_0 is generated by the vectors $v_0, v_1, \ldots, v_n$ in such a way that it contains the feasible region P. The points $v_1, \ldots, v_n$ are chosen on the rays emanating from v_0 through the neighbors of v_0. Note that by checking the coefficients of the simplex tableau that gives v_0 we can determine if it is possible to use one of the constraints $A_i x \leq b_i$ to truncate the cone formed by the n constraints binding at v_0. Otherwise, a vector $c \in R^n$ can be found such that v_0 is the unique solution of the linear program $\min_{x \in P} c^T x$. Then the problem $\max_{x \in P} c^T x$ is solved to obtain x_0. The intersection of the hyperplane $c^T x = c^T x_0$ with the n rays emanating from v_0 will generate $v_1, \ldots, v_n$.

The algorithm will generate a sequence of subproblems, with a corresponding sequence of bounded polyhedra $S_k \supseteq P$. These polyhedra will be characterized by a subset of constraints $Ax \leq b$, and the set of vertices of S_k. Let I_k denote the subset of $\{1, \ldots, m\}$ whose corresponding constraints define S_k, and V_k denote the set of vertices of S_k. The index set J_k denotes the subscripts of the elements in V_k.

Algorithm:

0. Let $V_0 = \{v_j : j \in J_0\}$ be the set of vertices of S_0 and let $u_0 = \min_{j \in J_0} \{f(v_j)\}$.

1. Given J_k and V_k solve the linear problem P_k:

$$\min_{\alpha} \sum_{j \in J_k} \alpha_j f(v_j)$$

$$\text{s.t.} \quad \sum_{j \in J_k} \alpha_j A v_j \leq b$$

$$\sum_{j \in J_k} \alpha_j = 1$$

$$\alpha_j \geq 0, \; j \in J_k$$

Let α^k be the solution with optimal function value $z_k = \sum_{j \in J_k} \alpha_j{}^k f(v_j)$. Set:

$$x^k = \sum_{j \in J_k} \alpha_j{}^k v_j$$

and let $\bar{J}_k = \{j : \alpha_j{}^k > 0, \; j \in J_k\}$. The set $\bar{J}_k$ identifies those vertices of S_k which are actually used in the representation of x^k.

2. (a) If $v_j \in P$ for each $j \in \bar{J}_k$, stop. The global solution is that $v^* \in V_k$ for which $f(v^*) = \min \{f(v_j) : j \in \bar{J}_k\}$.

(b) Otherwise , there is a $v_{j_k} \in V_k$ such that $\alpha_{j_k}{}^k > 0$ and v_{j_k} is not a vertex of the feasible domain P. Select any i_k not in I_k such that $A_{i_k} v_{j_k} > b_{i_k}$ and set $I_k = I_k \cup \{i_k\}$. (Here A_i denotes the ith row of A).

(c) Find the set of vertices V_{k+1} of S_{k+1} and update the bound u_k if any of the new vertices are vertices of P (that is $u_{k+1} = \min\{u_k, f(v_t)\}$ where v_t are new vertices in both S_{k+1} and P). Return to step 1.

The stopping rule of 2(a) above is justified by the next theorem.

Theorem 6.3.1: If all vertices v_j $(j \in \overline{J}_k)$ are in P, then the vertex v^* such that

$$f(v^*) = \min \{f(v_j):\ j \in \overline{J}_k\}$$

is a global solution of the original problem.

Proof: Let $F_k(x)$ be the convex envelope of f over S_k and x_k be as defined above. Then

$$F_k(x^k) = \sum_{j \in \overline{J}_k} \alpha_j{}^k f(v_j) \geq \min\{f(v_j):\ j \in \overline{J}_k\} = f(v^*) \geq$$

$$\min\{F_k(v_j):\ j \in \overline{J}_k\} \geq F_k(x^k)$$

and therefore $F_k(x^k) = f(v^*)$. But $F_k(x^k)$ underestimates $f(x)$ for all $x \in P$, so that $v^* \in P$ also yields a lower bound of f over P.

Note that the algorithm generates upper and lower bounds for the global minimum, that is

$$z_k \leq f(x^*) \leq u_k$$

The numbers u_k are upper bounds for the global optimum by the the way they are defined (use only function evaluations at vertices of P). The lower bounding property of z_k follows from the fact that problem P_k can be interpreted as finding the minimum over P of the convex envelope of f taken over $S_k \supseteq P$. This convex envelope underestimates f over P, so that its minimum over P is a lower bound to $f(v^*)$. In fact the lower bounds have the following monotonicity property:

Theorem 6.3.2: The sequence $\{z_k\}$, generated by the algorithm, is monotonically nondecreasing.

Proof: At step k the algorithm solves the problem of minimizing (over P) the convex envelope F_k of f taken over S_k. Since $S_{k+1} \subseteq S_k$, it follows that $F_{k+1}(x) \geq F_k(x)$ for all $x \in S^{k+1} \supseteq P$. Then

$$\min_{x \in P} F_{k+1}(x) \geq \min_{x \in P} F_k(x) \quad => \quad z_{k+1} \geq z_k .$$

Different computational issues of the algorithm can be found in detail in [HOFF76]. The algorithm described above has been coded and a number of computational results are reported. The method effectively uses the linear underestimating functions. The disadvantage of the algorithm is that the number of vertices may increase exponentially with iteration number and determining them is the most costly part of the algorithm. Therefore, even though the maximum number of rows involved in any subproblem is limited to the number of constraints in the original problem the number of columns may grow exponentially. A bound on the total number of subproblems that have to be generated is given by the number of constraints of the original problem. An extension of the algorithm for the case of general convex constraints is given by [HOFF81].

6.4 Techniques using special ordered sets

Beale and Tomlin ([BEAL76], [BEAL78] [BEAL79]), proposed a different method of the branch and bound type for global minimization. Their approach is based on the use of special ordered sets. Special ordered sets arise in the problem of approximating the separable objective function using piecewise linear approximations. If the function is convex then the piecewise approximation is equivalent to a linear problem. However, in the absence of convexity the piecewise linear approximation can be transformed to an equivalent zero-one linear program.

Special ordered sets are of two types [BEAL70]: Type $S1$ sets are groups of variables for which only one may take a nonzero value (multiple choices). Type $S2$ sets are groups of variables of which at most two adjacent members may take on nonzero values. The $S2$ concept has been considerably extended by Beale and

Forrest [BEAL76] to the linked ordered sets.

To understand the significance of special ordered sets consider the one dimensional case where the objective function is $f(x)$, $x \in R$. Suppose that the argument x can take only a finite number of possible values, say x_i, i=0, ..., m. Then we can introduce a set of nonnegative variables λ_i, i=0, ..., m such that

$$\sum_{i=1}^{m} \lambda_i = 1, \; x = \sum_{i=1}^{m} \lambda_i x_i$$

Then the nonlinear function $f(x)$ is represented by the linear function

$$\gamma(\lambda) = \sum_{i=1}^{m} \lambda_i f(x_i),$$

if we impose the additional restriction that no more than one of the λ_i's may be nonzero ($S1$ set). This approach is usually called the λ-formulation of separable programming. By solving the linear programming problem (without this additional restriction) we obtain a bound on the solution. Furthermore, in order to sharpen this bound we can carry out branch and bound operations on the set of λ-variables allowed to take nonzero values. Specifically, for any such λ_{i_0} note that

either $\lambda_0 = \cdots = \lambda_{i_0} = 0$

or $\lambda_{i_0+1} = \cdots \lambda_m = 0.$

So we can replace the original linear programming problem by two linear programming subproblems that include all valid solutions to the original problem while excluding the solution to the previous linear programming approximation. We can therefore solve the problem by a branch and bound method.

In the above formulation we may allow two adjacent λ_i's to be nonzero ($S2$ set). Geometrically, it amounts to permitting linear interpolation to $f(x)$ between adjacent x_i's. The modification of the branching rule is simply that

either $\lambda_0 = \cdots = \lambda_{i_0-1} = 0$

or $\lambda_{i_0+1} = \cdots = \lambda_m = 0.$

In general if we have the separable objective function $f(x) = \sum_{i=1}^{n} f_i(x_i)$, then we introduce n sets of nonnegative variables λ_{ij} for $i=1, \ldots, m_j$ and $j=1, \ldots, n$.

Define the new constraints

$$\sum_{i=1}^{m_j} \lambda_{ij} = 1, \quad j=1, \ldots, n$$

$$x_j = \sum_{i=1}^{m_j} \lambda_{ij} x_{ij}, \quad j=1, \ldots, n$$

and each nonlinear function $f_j(x_j)$ is represented by the linear function

$$\gamma_j(\lambda) = \sum_{i=1}^{m_j} \lambda_{ij} f_j(x_{ij}).$$

In [BEAL78] branch and bound procedures are used to find the global optimum of a function that is defined as a sum of products of functions of single arguments. These methods can be incorporated as extensions to integer programming facilities in general mathematical programming. Here the implementation uses special and linked ordered sets. A number of computational results are reported using SCICON's mathematical programming system SCICONIC. More discussion on branch and bound methods can be found in [BEAL80].

6.5 Factorable nonconvex programming

McCormick ([McCOR76], [McCOR83]) considers the global optimization problem for nonlinear programming problems which are factorable. A function $f(x)$, where $x \in R^n$, is called factorable if it can be represented as the last function in a finite sequence of functions $\{f_j(x)\}$ that are composed as follows:

$$f_j(x) = x^j, \quad j=1, \ldots, n.$$

For $j > n$, $f_j(x)$ is either of the form

$$f_k(x)+f_l(x), \quad k,l<j, \quad f_k(x)\times f_l(x), \quad k,l<j$$

or

$$T(\, f_k(x)\,), \quad k<j$$

where $T(f)$ is a function of a single variable. For nonlinear problems with a factorable objective function, McCormick proposes a procedure for obtaining tight underestimating convex problems. Important to this approach is the use of convex envelope of a function of a single variable.

McCormick's approach for obtaining a global solution to the nonlinear problem

$$\min_{x \in S} f(x)$$

is outlined next. Here $S \subseteq R^n$ is a bounded convex set. Let f^* denote the global minimum function value, and let $\varepsilon > 0$ be an acceptable tolerance for an approximate solution. Set $k = 1$ and $S^{1,1} = S$.

Iteration k $(k \geq 1)$

At the kth iteration we assume that there are k regions $S^{k,l}$, $(S^{k,l} \subseteq S$, $l=1, \ldots, k)$ defining k approximate problems

Problem $A^{k,l}$: $\min_{x \in S^{k,l}} f(x)$

with minimum function values $f^*_{k,l}$ such that

$$f^* = \min_{l} \{f^*_{k,l}\}$$

1: Obtain an approximate solution $x^{k,l}$ of Problem $A^{k,l}$, for $l=1, \ldots, k$.
2: Find a lower bound $f_{k,l}$ of $f^*_{k,l}$ for $l=1, ..., k$. Let $K=K(k)$ be the index such that

$$f_{k,K} = \min \{f_{k,1}, \ldots, f_{k,k}\}.$$

If $f(x_{k,l}) \leq f_{k,K}+\varepsilon$, then stop; $f(x_{k,l})$ is an acceptable ε-approximate solution to the initial problem.

Otherwise go to the $(k+1)$th iteration. The approximate problems $A^{k+1,l}$ are identical to $A^{k,l}$ for $l \neq K$ and the two new problems $A^{k+1,K}$ and $A^{k+1,k+1}$ have constraint sets $S^{k+1,K}$ and $S^{k+1,k+1}$ respectively, where these two constraint sets form a partition of $S^{k,K}$, that is

$$S^{k+1,K} \cup S^{k+1,k+1} = S^{k,K}.$$

From the practical point of view, it is important to solve efficiently the subproblems $A^{k,l}$. In case the objective function is factorable, convex underestimating functions can be considered. Procedures for creating such convex underestimators are discussed in [McCO83]. Also other techniques for lower bound estimators can be used. Under certain conditions convergence to the global minimum is obtained. The efficiency of these methods depends upon the tightness of the underestimating problems.

6.6 Exercises

1. Consider the following separable concave programming problem:

$$\min_x f(x) = \sum_{i=1}^{n} f_i(x_i)$$

$$\text{s.t. } Ax \leq b\,,\, 0 \leq x_i \leq \beta_i\,,\; i = 1, \ldots, n$$

where each function $f_i(x_i)$ is concave for $x \in [0, \beta_i]$.
Partition each interval $[0, \beta_i]$ into k_i subintervals of length $h_i = \beta_i / k_i$, and introduce the new variables $0 \leq \omega_{ij} \leq 1$, such that

$$x_i = h_i \sum_{j=1}^{k_i} \omega_{ij},\; i=1, \ldots, n.$$

Using these new variables find the piecewise linear function $\Gamma_i(\omega)$ that interpolates $f_i(x_i)$ at $x_i = jh_i$, $j=0,1, \ldots, k_i$. Consider the piecewise linear (concave) programming problem

$$\min_\omega \Gamma(w) = \sum_{i=1}^{n} \Gamma_i(\omega)$$

over the transformed constrained set. Formulate this problem as a mixed zero-one integer program with a linear objective function.

2. Consider the following problem

$$\min f(x, y) = -(x-20)^2-(y-10)^2$$

$$\text{s.t } 2y-x \le 20,\ x^2-(y-10)^2 \le 500,\ x,y \ge 0$$

Use Horst's algorithm to prove that $(x,y) = (0,0)$ is the global solution.

6.7 References

[BENS82A] Benson, H.P. *On the convergence of two branch and bound algorithms for nonconvex programming problems.* JOTA 36 (1982), 129-134.

[BENS82B] Benson, H.P. *Algorithms for parametric nonconvex programming.* JOTA 38 (1982), 316-340.

[BENS85] Benson, H.P. *A finite algorithm for concave minimization over a polyhedron.* Naval. Res. Log. Quarterly 32 (1985), 165-177.

[BEAL80] Beale, E.M. *Branch and bound methods for numerical optimization of nonconvex functions.* Proc. Fourth Sympos. on Computational Statistics, Edinburgh 1980, pp 11-20, Physica, Vienna.

[BEAL76] Beale, E.M. and Forrest, J.J.H. *Global optimization using special ordered sets.* Math. Progr. 10 (1976), 52-69.

[BEAL78] Beale, E.M. and Forrest, J.J.H. *Global optimization as an extension of integer programming.* In: Towards Global Optimization 2, L.C. Dixon and G.P. Szego, eds. North-Holland, Amsterdam (1978), 131-149.

[BEAL70] Beale, E.M. and Tomlin, J.A. *Special facilities in a general mathematical programming system for nonconvex problems using ordered sets of variables.* Proc. 5th Intern. Confer. on Oper. Research, 1970 (ed. J. Lawrence), 447-454.

[BEAL79] Beale, E.M. *Branch and bound methods for mathematical programming systems.* Annals of Discrete Math. 5 (1979), 201-210.

[FALK72] Falk, J.E. *An algorithm for locating approximate global solutions for nonconvex separable problems.* Technical Paper Serial T-262 (1972), Program in Logistics, George Washington Univ.

[FALK69] Falk, J.E. and Soland, R.M. *An algorithm for separable nonconvex programming problems*. Management Sc. 15 (1969), 550-569.

[FALK76] Falk, J.E. and Hoffman K.L. *A successive underestimating method for concave minimization problems*. Math. Oper. Reser. 1 (1976), 251-259.

[GROT76] Grotte, J.H. *Program MOGG-A code for solving separable nonconvex optimization problems*. P-1318 (1976), The Institute for Defense Analysis, Arlington, VA.

[HOFF75] Hoffman, K.L. *A successive underestimating method for concave minimization problems*. Ph.D. thesis (1975), George Washington Univ.

[HOFF81] Hoffman, K.L. *A method for globally minimizing concave functions over convex sets*. Math. Progr. 20 (1981), 22-32.

[HORS76] Horst, R. *An algorithm for nonconvex programming problems*. Math. Progr. 10 (1976), 312-321.

[HORS78] Horst, R. *A new approach for separable nonconvex minimization problems including a method for finding the global minimum of a function of a single variable*. Proceedings in Oper. Res. 7, Sixth Annual Meeting, Deutsch-Geselsch. Oper. Res. Christian-Albrechts, Univ. Kiel 1978, 39-47.

[HORS80] Horst, R. *A note on the convergence of an algorithm for nonconvex programming problems*. Math. Progr. 19 (1980), 237-238.

[HORS85] Horst, R. *A general class of branch and bound methods in global optimization with some new approaches for concave minimization*. To appear in JOTA (1985).

[McCO72] McCormick, G.P. *Attempts to calculate global solutions of problems that may have many local minima*. Numerical Methods in Nonlinear Optimization, F.A. Lootsma, ed. Academic Press, N.Y. (1972), 209-221.

[McCO76] McCormick, G.P. *Computability of global solutions to factorable nonconvex programs: Part I- Convex underestimating problems*. Math. Progr. 10 (1976), 147-175.

[McCO83] McCormick, G.P. *Nonlinear Programming: Theory algorithms and applications*. John Wiley and Sons, N.Y. (1983).

[SOLA71] Soland, R.M. *An algorithm for separable nonconvex programming problems II: nonconvex constraints*. Management Sci. 17 (1971), 759-773.

Chapter 7 Bilinear Programming methods for nonconvex quadratic problems

7.1 Introduction

The bilinear programming problem belongs to the class of nonconvex optimization problems and in its general form can be stated as follows:

$$\text{global } \min\{f(x, y): x \in X, y \in Y\} \tag{1}$$

where X, Y are polyhedral sets in R^n and R^m respectively and $f(x,y)$ is a bilinear function. The difficulty of the problem is that it may have many local optima. We restrict our attention here to the case of quadratic objective functions, i.e. problems of the form

$$\text{global } \min_{x,y} f(x,y) = c^T x + d^T y + x^T Q y \tag{2}$$

such that

$$x \in X = \{x \in R^n : A_1 x = b_1, x \geq 0\}$$

$$y \in Y = \{y \in R^m : A_2 y = b_2, y \geq 0\}$$

where Q, A_1, A_2 are matrices of dimensions $n \times m$, $k \times n$, $l \times m$, respectively, and $c \in R^n$, $d \in R^m$, $b_1 \in R^k$, $b_2 \in R^l$.

The above problem can be easily expressed as an indefinite quadratic program of special structure. Let

$$z = \begin{bmatrix} x \\ y \end{bmatrix}, \ a = \begin{bmatrix} c \\ d \end{bmatrix}, \ M = \begin{bmatrix} 0 & Q \\ Q^T & 0 \end{bmatrix}, \ A = \begin{bmatrix} A_1 & 0 \\ 0 & A_2 \end{bmatrix}, \ b = \begin{bmatrix} b_1 \\ b_2 \end{bmatrix}.$$

Then the quadratic bilinear program can be written in the following form:

$$\text{global } \min_z a^T z + \frac{1}{2} z^T M z \tag{3}$$

$$\text{s.t. } z \in Z = \{z \in R^{n+m} : Az = b, z \geq 0\}.$$

In the above formulation problem (3) has an objective function which is neither convex nor concave. However, we can prove that the general bilinear problem (1) is equivalent to the concave minimization problem under linear constraints.

Assume that $\min_{y \in Y} f(x,y)$ exists for every $x \in X$ and X, Y are bounded and nonempty. Then we have

Theorem 7.1.1: Problem (1) can be formulated as a concave piecewise linear minimization problem.

Proof: Let $V(Y)$ denote the set of vertices of Y. Then we have

$$\min_{x \in X, y \in Y} f(x,y) = \min_{x \in X}\{\min_{y \in Y} f(x,y)\}$$

$$= \min_{x \in X}\{\min_{y \in V(Y)} f(x,y)\} = \min_{x \in X} g(x)$$

where $g(x) = \min_{y \in V(Y)} f(x,y) = \min_{y \in Y} f(x,y)$. Since $V(Y)$ is a finite set, $g(x)$ is a concave piecewise linear function of x.

Also it can be shown that a concave piecewise linear minimization problem can be formulated as a bilinear program.

Theorem 7.1.2: Let $g(x) = \min_{1 \le i \le m} \{l_i(x)\}$ be a concave piecewise linear function defined as the pointwise minimum of m linear functions $l_i(x)$. Then $\min_{x \in X} g(x)$, where X is a compact convex set in R^n, can be formulated as a bilinear problem.

Proof: Consider the following bilinear function

$$f(x,y) = \sum_{i=1}^{m} y_i l_i(x), \quad x \in R^n, \, y \in R^m$$

Define the $(m-1)$-simplex in R^m,

$$S = \{y \in R^m : \sum_{i=1}^{m} y_i = 1 \quad y \ge 0\},$$

whose vertices are $v_1, \ldots, v_m$ with v_i the ith unit vector in R^m. Note that $f(x,v_i) = l_i(x)$, $i=1, \ldots, m$. Hence

$$\min_{y \in Y} f(x,y) = \min_{1 \le i \le m} f(x,v_i) = \min_{1 \le i \le m} l_i(x) = g(x) \quad (x \in R^n)$$

Therefore,

$$\min_{x\in X} g(x) = \min_{x\in X} \{\min_{y\in Y} f(x,y)\} = \min_{x\in X, y\in Y} f(x,y)$$

Since every concave function can be approximated arbitrarily close by piecewise linear concave functions, the general concave problem is equivalent to a bilinear programming problem [THIE80].

7.2 Maximization of convex quadratic functions

A number of authors recognized the relationship between algorithms for bilinear programming and algorithms for convex maximization (or concave minimization). Therefore specialized bilinear programming algorithms can be adapted for quadratic convex programming.

Konno [KONN76B], considers the following quadratic program:

$$\max f(x) = 2c^T x + x^T Qx \tag{4}$$

$$\text{s.t. } Ax = b, \ x \geq 0$$

where $c, x \in R^n$, $b \in R^m$, A is an $m \times n$ matrix and Q is an $n \times n$ symmetric positive semidefinite matrix. Assume that the feasible region is nonempty and bounded. Consider now the bilinear program

$$\max \phi(x,y) = c^T x + c^T y + x^T Qy \tag{5}$$

$$\text{s.t. } Ax = b, Ay = b, \ x, y \geq 0.$$

Theorem 7.2.1:

a) If problem (5) has an optimal solution (v^*, u^*) then v^* and u^* are both optimal solutions for problem (4).

b) If x^* is an optimal solution for (4), then $(x,y) = (x^*, x^*)$ is also an optimal solution for (5).

Proof: a) Let $X = \{x : Ax = b, x \geq 0\}$. First note that $f(x^*) \geq f(x)$ for all $x \in X$ and in particular $f(x^*) \geq f(v^*) = \phi(v^*, v^*)$ and $f(x^*) \geq f(u^*) = \phi(u^*, u^*)$. Also

$$\phi(v^*,u^*) = \max_{x \in X, y \in X} \phi(x,y) \geq \max_{x \in X} \phi(x,x) = f(x^*).$$

b) Since $f(v^*), f(u^*) \geq f(x^*)$ and $\phi(x^*,x^*) = f(x^*) = \phi(v^*,u^*)$ we need only to prove that $\phi(v^*,u^*) = f(v^*) = f(u^*)$. To prove this note that since (v^*,u^*) is optimal for (5), we have

$$0 \leq \phi(v^*,u^*) - \phi(v^*,v^*) = c^T(u^*-v^*) + (v^*)^T Q(u^*-v^*)$$

$$0 \leq \phi(v^*,u^*) - \phi(u^*,u^*) = c^T(v^*-u^*) + (u^*)^T Q(v^*-u^*).$$

Adding the last two inequalities we have

$$(v^*-u^*)^T Q(v^*-u^*) \leq 0$$

and since Q is positive semidefinite we must have $Q(v^*-u^*) = 0$. Then the above inequalities give $c^T(v^*-u^*) = 0$ and hence $\phi(v^*,v^*) = \phi(u^*,u^*)$.

The cutting plane method for bilinear programming developed by Konno [KONN76A] (see also [KONN71A, B]) is adapted for this problem. An iterative procedure is developed that improves cuts by exploiting the symmetric structure of bilinear problems. Related problems and methods are studied in [KONN80] and [KONN81].

Czochralska [CZOC82A], considers the quadratic bilinear problem of the form (2), where the feasible domain may be unbounded. The bilinear programming algorithm proposed here uses linear programming and methods of ranking the extreme points of convex polyhedra. The method searches for an optimal solution among those basic solutions which are equilibrium points of the problem. A feasible solution (x_0,y_0) is called an equilibrium point of (2) if

$$f(x_0,y_0) = \min_{x \in X} f(x,y_0) = \min_{x \in Y} f(x_0,y).$$

Examples illustrating this algorithm and numerous theoretical results concerning bilinear programs are given in [CZOC82A]. Based on this algorithm, in [CZOC82B] an algorithm is given for nonconvex quadratic problems. This bilinear programming algorithm for quadratic programming problems is greatly simplified as a consequence of the problem structure and because of the fact that the verification of necessary and sufficient conditions for the existence of an optimal solution is reduced to the solution of a linear programming problem.

Some other approaches for bilinear programming are described in [SHER80], [VAIS76], [VAIS77], [ALTM68] and [GALL77]. The algorithm described in the last paper uses a minimax formulation by developing a cutting plane approach which is similar to Tuy's method for concave programming.

7.3 Jointly constrained bilinear problems

The traditional bilinear problem we discussed in the previous section has the important property that its solution occurs at a vertex point. This property is lost when we consider the more general bilinear problem with joint constraints. Such problems have been considered in [AL-K83] and have the form

$$\min_{(x,y)} \phi(x,y) = f(x)+x^T y+g(y) \tag{6}$$

$$\text{s.t.} \quad (x,y) \in S \cap \Omega$$

where f, g are convex, S is a closed and convex set, and Ω is a rectangle given by

$$\Omega = \{(x,y)\colon l \le x \le L,\ m \le y \le M\}.$$

An algorithm of the branch and bound type is proposed, using convex envelopes, which is guaranteed to converge to a global solution. The convex envelopes that are used in this approach are given by the following theorem:

Theorem 7.3.1: The convex envelope $E(x,y)$ of the function $x^T y$ over Ω is given by

$$E(x,y) = \sum_{i=1}^{n} E_i(x_i,y_i)$$

where $E_i(x_i,y_i)$ is the convex envelope of $x_i y_i$ in

$$\Omega_i = \{(x_i,y_i)\colon l_i \le x_i \le L_i,\ m_i \le y_i \le M_i\} \subseteq R^2$$

Proof: It is relatively easy to prove that

$$E_i(x_i,y_i) = \max\{m_i x_i + l_i y_i - l_i m_i,\ M_i x_i + L_i y_i - L_i M_i\}.$$

Also note that $E_i(x_i,y_i) = x_i y_i$ for all (x_i,y_i) on the boundary of Ω_i.

Since the functions f, g are convex in (6), the convex envelope of the objective function is give by

$$\Psi^1 = \Psi(x,y) = f(x) + E(x,y) + g(y)$$

Also note that $\Psi(x,y) \le \phi(x,y)$ (with equality on the boundary of Ω). The branch-and-bound algorithm is described next.

At step k the rectangle Ω will be partitioned into a set of subrectangles

$$\Omega^{kj} = \{(x,y): l^{kj} \le x \le L^{kj},\ m^{kj} \le y \le M^{kj}\}.$$

The convex envelope of $\phi(x,y)$ over Ω^{kj} is called Ψ^{kj}. Define now the function

$$\Psi^k(x,y) = \Psi^{kj}(x,y) \ \text{ if } \ (x,y) \in \Omega^{kj}$$

This is a well-defined function for all $(x,y) \in \Omega$, since each of the functions Ψ^{kj} agree with ϕ along the boundary of Ω^{kj}. The solution (x^k,y^k) of the convex underestimating problem

Problem P^k: $\min\{\Psi^k(x,y): (x,y) \in S \cap \Omega\}$
gives a lower bound $v^k = \Psi^k(x^k,y^k)$ of the global optimum ϕ^*. Also $V^k = \phi(x^*,y^*)$ gives an upper bound on ϕ^* since $(x^*,y^*) \in S \cap \Omega$.

With each convex envelope Ψ^{kj} and Ω^{kj} is associated the convex problem

Problem P^{kj}: $\min\{\Psi^{kj}(x,y): (x,y) \in S \cap \Omega^{kj}\}$
If $v^{kj} = \Psi^{kj}(x^{kj},y^{kj})$ is the solution of P^{kj}, then

$$v^k = \min_j \{v^{kj}\}$$

and (x^k,y^k) is any of the points (x^{kj},y^{kj}) which gives the value v^k.

Initially, step 1 consists of Problem P^1, with Ψ^1 defined above. Solving P^1 we obtain (x^1,y^1) and lower and upper bounds

$$v^1 \le \phi^* \le V^1.$$

If we have equality above we stop; (x^1,y^1) is optimal. If $v^1 < V^1$, then we must have

$$E(x^1,y^1) < (x^1)^T y^1$$

or equivalently

$$\max\{m_i x_i^1 + l_i y_i^1 - l_i m_i,\ M_i x_i^1 + L_i y_i^1 - L_i M_i\} < x_i^1 y_i^1$$

for some i. Use this inequality to split Ω. Choose an I which produces the largest difference between the two sides of the above inequality, and split the I rectangle into four subrectangles, creating the new subproblems $P^{21}, \ldots, P^{24}$ for step 2. By a similar argument we move from step k to step $k+1$ of the algorithm. Details with a convergence proof and examples are given in [AL-K83]. From the computational point of view we should be concerned here with the infinite convergence property of the algorithm. However in practice we are satisfied if an ε-approximate solution is obtained with a reasonable computational effort.

7.4 Exercises

1. Consider the bilinear function $F(x,y) = c^T x + c^T y + x^T Q y$, defined on $P \times P$, ($P = \{x : Ax = b,\ x \geq 0\}$ is a bounded polyhedron), where Q is an $n \times n$ symmetric positive semidefinite matrix, A is an $m \times n$ matrix, c, $x \in R^n$ and $b \in R^m$. Then prove that
   a) $F(x,y) \leq \max\{F(x,x),\ F(y,y)\}$ for all $(x,y) \in P \times P$.
   b) If Q is positive definite and $x \neq y$, the above inequality is strict.

2. Solve the bilinear programming problem:

$$\min_{(x,y)} -x + xy - y$$

$$\text{s.t.}\quad -6x + 8y \leq 3,\ 3x - y \leq 3$$

$$0 \leq x,\ y \leq 5$$

   Does this problem have any extreme point solution?

3. The general jointly concave bilinear programming problem has the form:

$$\min_{(x,y) \in S} \quad F(x,y) = f(x) + x^T y + g(y)$$

   where S is a compact convex set in R^{2n} and $f(x)$, $g(y)$ are continuous concave functions on S. Prove that the problem has a solution that occurs on the boundary of S.

7.5 References

[AL-K83] Al-Khayyal, F. and Falk, J.E. *Jointly constrained biconvex programming.* Math. of Oper. Res., Vol. 8, No. 2 (1983), 273-286.

[ALTM68] Altman, M. *Bilinear programming.* Bull. Acad. Pol.Sci. Ser. Sci. Math. Astronom. Phy. 19 (1968), 741-746.

[CZOC82A] Czochralska, I. *Bilinear programming.* Zastosowania Mathematyki XVII, 3 (1982), 495-514.

[CZOC82B] Czochralska, I. *The method of bilinear programming for nonconvex quadratic programming.* Zastosowania Mathematyki XVII, 3 (1982), 515-525.

[GALL77] Gallo, G. and Ulkulu, A. *Bilinear programming: An exact algorithm.* Mathem. Progr. 12 (1971), 173-194.

[KONN71A] Konno, H. *Bilinear programming: Part I. An algorithm for solving bilinear programs.* Technical Report 71-9 (1971), Oper. Res., Stanford Univ.

[KONN71B] Konno, H. *Bilinear programming: Part II. Applications of bilinear programming.* Technical Report 71-10 (1971), Oper. Res., Stanford Univ.

[KONN76A] Konno, H. *A cutting plane algorithm for solving bilinear programs.* Math. Progr. 11 (1976), 14-27.

[KONN76B] Konno, H. *Maximization of a convex quadratic function under linear constraints.* Math. Progr. 11 (1976), 117-127.

[KONN80] Konno, H. *Maximizing a convex function over a hypercube.* J. Oper. Res. Soc. Japan 23 (1980), 171-189.

[KONN81] Konno, H. *An algorithm for solving bilinear knapsack problems.* J. Oper. Res. Soc. Japan 24 (9181), 360-373.

[SHER80] Sherali, H. and Shetty, C.M. *A finitely convergent algorithm for bilinear programming problems using polar cuts and disjunctive face cuts.* Math. Progr. 19 (1980), 14-31.

[THIE80] Thieu, T.V. *Relationship between bilinear programming and concave minimization under linear constraints.* Acta Math. Vietnam 5 (1980), 106-113.

[VAIS76] Vaish, H. and Shetty C.M. *The bilinear programming problem.* Naval Res. Logist. Quarterly 23 (1976), 303-309.

[VAIS77] Vaish, H. and Shetty C.M. *A cutting plane algorithm for the bilinear programming problem.* Naval Res. Logist. Quarterly 24 (1977), 83-94.

Chapter 8 Large scale problems

8.1 Concave quadratic problems

In this section we consider large scale problems which may include many variables that appear only linearly. Specifically, we look at the following type of problem:

$$\text{global } \min \Psi(z,y) \qquad \text{(P)}$$

$$\text{s.t. } (z,y) \in \Omega = \{(z,y): A_1 z + A_2 y = b,\ z \geq 0,\ y \geq 0\} \subseteq R^{n+k}$$

where $\Psi(z,y) = \phi(z) + d^T y$, and $\phi(z)$ is a concave quadratic function, given by

$$\phi(z) = c^T z - \frac{1}{2} z^T Q z$$

with Q a positive semi-definite symmetric matrix. We assume that $z \in R^n$, $y \in R^k$, $A_1 \in R^{m \times n}$, $A_2 \in R^{m \times k}$, and that the polyhedral set Ω is nonempty and compact. We denote by Ψ^* the global minimum function value. In previous sections only the case corresponding to $k=0$ has been considered, that is, all variables were treated in the same manner. A variety of important applications lead to such a large scale formulation, including plant location with economies of scale, quadratic assignment problems and fixed charge problems [PARD86]

The motivation for considering this type of problem is similar to that for problems of large scale convex minimization or zero-one integer linear problems. Large scale, linearly constrained minimization problems with a convex objective function can be solved provided that most of the variables appear linearly on the objective function. In that case a method such as MINOS [MURT78], [MURT83], which treats the linear variables in a different manner than those appearing nonlinearly can be used very efficiently. Similarly large scale zero-one mixed integer linear programming can be solved in a reasonable time provided most of the variables are continous.

The approach described here takes full advantage of the linearity of y variables [ROSE86]. Some of the ideas of this new approach were proposed originally by [ROSE83] where a computational method is given for concave quadratic programming. A similar approach is used for the large scale problem. Initially a rectangular domain $R_z \subseteq R^n$ is determined, which contains the projection Ω_z of Ω on the z-space. This is done by solving a multiple-cost-row linear program with $2n$ objective functions. Then a linear underestimating function $\Gamma(z)$ to $\phi(z)$ is easily computed and the following linear problem is solved,

$$\min_{(z,y)\in\Omega} \quad \Gamma(z)+d^T y$$

to obtain upper and lower bounds for the global minimum Ψ^*. Computational algorithms based on this initial step, but using different methods for constructing underestimating problems, have been developed by [KALA84] and [ZILV83].

8.2 Separable formulation

In this approach we first show how the original large scale problem can be reduced to a separable quadratic problem using the eigenstructure of the quadratic form. When the problem has a separable objective function we can easily obtain linear and piecewise linear approximations [PARD85].

To carry out the reduction we note that the matrix Q in problem (P) is a positive semi-definite symmetric matrix. Therefore we can compute the (real) eigenvalues $\lambda_1,\lambda_2, \ldots ,\lambda_n$ of Q and the corresponding (orthonormal) eigenvectors $u_1,u_2, \ldots ,u_n$. Then $Q = UDU^T$ where $U = [u_1, \ldots ,u_n]$, $U^T U = I$ and $D = diag(\lambda_1, \ldots ,\lambda_n)$.

Consider the linear transformation $z = U\hat{x}$, or equivalently $\hat{x}_i = u_i^T z$. Initially solve the multiple-cost-row linear program with $2n$ rows:

$$\min u_i^T z \qquad \text{(MCR)}$$

$$-\min -u_i^T z$$

$$\text{s.t. } (z,y) \in \Omega, \; i = 1, 2,\ldots, n$$

Denote by $\underline{x}_i$ and $\overline{x}_i$ the optimal function values corresponding to u_i, and $-u_i$ respectively, and let $x_i = \hat{x}_i - \underline{x}_i$, and $\beta_i = \overline{x}_i - \underline{x}_i$. Now the original problem (P) can be formulated as a separable programming problem in terms of the new variables x_i

$$\min \sum_{i=1}^{n} q_i(x_i) + d^T y + q_0$$

$$\text{s.t. } A_3 x + A_2 y = \underline{b}$$

$$Ux \geq -U\underline{x}, \; 0 \leq x_i \leq \beta_i, \; y \geq 0$$

where

$$q_i(x_i) = \alpha_i x_i - \frac{1}{2}\lambda_i x_i^2, \; \alpha_i = c^T u_i - \lambda_i \underline{x}_i \;, \underline{x}^T = (\underline{x}_1, \ldots, \underline{x}_n),$$

$$q_0 = \sum_{i=1}^{n} (c^T u_i - \frac{1}{2}\lambda_i \underline{x}_i)\underline{x}_i \text{ and } A_3 = A_1 U \;, \underline{b} = b - A_3 \underline{x}.$$

From now on, for simplicity of notation, we consider the following separable concave quadratic problem:

$$\text{global min} \sum_{i=1}^{n} q_i(x_i) + d^T y \qquad \text{(SP)}$$

$$\text{s.t. } (x,y) \in \Omega = \{(x,y) \colon A_1 x + A_2 y = b, \, x \geq 0, \, y \geq 0\}$$

where $q_i(x_i) = c_i x_i - \frac{1}{2}\lambda_i x_i^2$. We also let $\phi(x) = \sum_{i=1}^{n} q_i(x_i)$, and $\Psi(x,y) = \phi(x) + d^T y$.

8.3 Linear Underestimator and Error Bounds

Based on the reduction described in the previous section we now consider the separable problem. The known values of β_i allow us to construct the smallest rectangular domain R_x in the x-space which contains Ω_x, the projection of Ω on the x-space. A linear function $\Gamma(x)$ is also easily obtained, which interpolates $\phi(x)$ at every vertex of R_x, and underestimates $\phi(x)$ on R_x, and therefore on the polytope

Ω_x. The rectangular domain is given by

$$R_x = \{x : 0 \le x_i \le \beta_i , i=1,...,n\}$$

A discussion of the equivalent construction without making explicit reduction to a separable problem is given in [ROSE85].

The linear underestimating function $\Gamma(x)$ is given by

$$\Gamma(x) = \sum_{i=1}^{n} \gamma_i(x_i), \quad \text{where} \quad \gamma_i(x_i) = (c_i - \frac{1}{2}\lambda_i \beta_i) x_i$$

The following linear program is then solved

$$\min_{(x,y)} \Gamma(x) + d^T y \qquad \text{(LU)}$$

$$\text{s.t.} \quad (x,y) \in \Omega$$

Note that this differs from (MCR) only in its objective function. The solution to this problem will give a vertex $\hat{v} = (\hat{x},\hat{y})$ of Ω which is a candidate for the global minimum Ψ^* of the original problem (SP). Since $(\hat{x},\hat{y})$ is feasible and also minimizes $\Gamma(x) + d^T y$ on Ω, it follows that

$$\Gamma(\hat{x}) + d^T \hat{y} \le \Psi^* \le \phi(\hat{x}) + d^T \hat{y} = \Psi(\hat{x},\hat{y})$$

Define the difference

$$E(x) = \phi(x) - \Gamma(x) \qquad (1)$$

The error at $(\hat{x},\hat{y})$ is given by $\Psi(\hat{x},\hat{y}) - \Psi^*$, and this error is bounded by

$$\Psi(\hat{x},\hat{y}) - \Psi^* \le E(\hat{x}) \qquad (2)$$

If $E(\hat{x})$ is sufficiently small, we consider $\Psi(\hat{x},\hat{y})$ to be an acceptable approximation to the global minimum Ψ^* with $(\hat{x},\hat{y})$ the corresponding approximate global minimum vertex.

We can now obtain a bound on $E(\hat{x})$. We do this relative to the range of $\phi(x)$ over R_x.

Let $\phi_{max} = \max_{x \in R_x} \phi(x)$, and $\phi_{min} = \min_{x \in R_x} \phi(x)$

Then we use $\Delta\phi_m = \phi_{max} - \phi_{min}$ as a scaling factor to measure $E(x)$ on Ω.

We first need a lower bound on $\Delta\phi$ which is given by the lemma to follow. Without loss of generality we can assume that

$$\lambda_1\beta_1^2 \geq \lambda_i\beta_i^2 \quad i = 2, \ldots, n \tag{3}$$

Define the ratios

$$\rho_i = \frac{\lambda_i\beta_i^2}{\lambda_1\beta_1^2} \leq 1 \quad i=1, 2, \ldots, n \tag{4}$$

The unconstrained maximum of $q_i(x_i)$ is attained at the point $\overline{x_i} = c_i/\lambda_i, i=1, \ldots, n$. The lower bound for $\Delta\phi_m$ depends on the distance between $\overline{x_i}$ and the midpoint of $[0,\beta_i]$ for $\overline{x_i} \in [0,\beta_i]$, and is independent of $\overline{x_i}$ for $\overline{x_i} \notin [0,\beta_i]$. We represent this dependence in terms of the quantities

$$\eta_i = \min\{1, |\frac{2\overline{x_i}}{\beta_i} - 1|\}, i=1, \ldots, n \tag{5}$$

Note that $0 \leq \eta_i \leq 1$.

Lemma 8.3.1:

$$\Delta\phi_m \geq \frac{1}{8}\lambda_1\beta_1^2 \sum_{i=1}^{n} \rho_i(1 + \eta_i)^2$$

Proof:

Let $\max q_i = \max_{0 \leq x_i \leq \beta_i} q_i(x_i)$, $\min q_i = \min_{0 \leq x_i \leq \beta_i} q_i(x_i)$ and $\Delta q_i = \max q_i - \min q_i$.

There are four cases to consider.

i) $\overline{x_i} \in [0,\beta_i/2]$: We have $q_i(\beta_i) = \lambda_i\beta_i(\overline{x_i} - \frac{1}{2}\beta_i) \leq 0$ so that $\min q_i = q_i(\beta_i)$ and $\max q_i = \frac{1}{2}\lambda_i\overline{x_i}^2$. From (5) we have $\overline{x_i} = \frac{1}{2}\beta_i(1-\eta_i)$. Then

$$\Delta q_i = \frac{1}{2}\lambda_i \bar{x}_i^2 - \lambda_i \beta_i \bar{x}_i + \frac{1}{2}\lambda_i \beta_i^2 = \frac{1}{2}\lambda_i (\beta_i - \bar{x}_i)^2$$
$$= \frac{1}{8}\lambda_i \beta_i^2 (1 + \eta_i)^2$$

ii) $\bar{x}_i \in [\beta_i/2, \beta_i]$: Now $q_i(\beta_i) \geq 0$, so that min $q_i = q_i(0) = 0$

From (5) we have $\bar{x}_i = \frac{1}{2}\beta_i(1 + \eta_i)$, so that $\Delta q_i = \frac{1}{2}\lambda_i \bar{x}_i^2 = \frac{1}{8}\lambda_i \beta_i^2 (1 + \eta_i)^2$.

iii) $\bar{x}_i \leq 0$: For $\bar{x}_i \leq 0$, we must have $c_i \leq 0$. Then $q_i(x_i) = -\frac{1}{2}\lambda_i x_i^2$, so that $\max q_i = 0$ and min $q_i \leq -\frac{1}{2}\lambda_i \beta_i^2$

Therefore since $\eta_i = 1$, $\Delta q_i \geq \frac{1}{2}\lambda_i \beta_i^2 = \frac{1}{8}\lambda_i \beta_i^2 (1 + \eta_i)^2$.

iv) $\bar{x}_i \geq \beta_i$: Now $\min q_i = q_i(0) = 0$, and since $\eta_i = 1$,

$\Delta q_i = \max q_i = q_i(\beta_i) = \lambda_i \beta_i (\bar{x}_i - \beta_i/2) \geq \frac{1}{2}\lambda_i \beta_i^2 = \frac{1}{8}\lambda_i \beta_i^2 (1 + \eta_i)^2$

Finally we have $\Delta\phi_m = \sum_{i=1}^{n} \Delta q_i \geq \frac{1}{8} \sum_{i=1}^{n} \lambda_i \beta_i^{\,2} (1+\eta_i)^2$.

An upper bound on $E(x)$ is easily obtained. We have

$$E(x) = \phi(x) - \Gamma(x) = \sum_{i=1}^{n} [q_i(x_i) - \gamma_i(x_i)] = \frac{1}{2}\sum_{i=1}^{n} \lambda_i (\beta_i - x_i) x_i$$

This attains its maximum at $x_i = \beta_i/2 \;\; i=1,...n$, so that for any $x \in \Omega_x \subset R_x$, we have:

$$E(x) \leq \frac{1}{8} \sum_{i=1}^{n} \lambda_i \beta_i^2 = \frac{1}{8} \lambda_1 \beta_1^2 \sum_{i=1}^{n} \rho_i \qquad (6)$$

Recall that $(\hat{x}, \hat{y})$ is the vertex obtained by solving the problem

$$\min \Gamma(x) + d^T y$$
$$(x, y) \in \Omega$$

Theorem 8.3.1: An a-priori bound on the relative error in $\Psi(\hat{x},\hat{y})$ is given by

$$\frac{\Psi(\hat{x},\hat{y}) - \Psi^*}{\Delta\phi_m} \le \frac{\sum_{i=1}^{n}\rho_i}{\sum_{i=1}^{n}\rho_i(1+\eta_i)^2} \equiv \sigma(\rho,\eta) \tag{7}$$

Proof: Use the lemma and the fact that $\Psi(\hat{x},\hat{y}) - \Psi^* \le E(\hat{x})$.

Note that $\sigma(\rho,\eta) \in [\frac{1}{4},1]$ and furthermore that $\sigma(\rho,\eta) < 1$, unless $\overline{x}_i = \beta_i/2$ for every i. In particular, if $\overline{x}_i \notin (0,\beta_i)$ for $i=1,\ldots,n$, we have $\sigma(\rho,\eta) = \frac{1}{4}$. This situation is characterized by the fact that $\max_{x\in R_x} \phi(x)$ occurs at any vertex of R_x.

8.4 Piecewise Linear Approximation and Zero-one Integer Formulation.

Using a piecewise linear approximation to each function $q_i(x_i)$ we can formulate a mixed integer zero-one LP, such that we can guarantee finding a solution $(\hat{x},\hat{y})$ for which

$$\frac{\Psi(\hat{x},\hat{y}) - \Psi^*}{\Delta\phi_m} \le \frac{E(\hat{x})}{\Delta\phi_m} \le \varepsilon \tag{8}$$

for any specified tolerance ε.
We partition each interval $[0,\beta_i]$ into k_i equal subintervals of length $h_i = \beta_i/k_i$, and introduce new variables ω_{ij}, such that

$$x_i = h_i \sum_{j=1}^{k_i} \omega_{ij}, \quad i=1,\ldots,n \tag{9}$$

The variables ω_{ij} are restricted to $\omega_{ij} \in [0,1]$, and furthermore the vector $\overline{\omega}_i = (\omega_{i1},\omega_{i2},\ldots,\omega_{ik_i})$ is restricted to have the form $\overline{\omega}_i = (1,\ldots,1,\omega_{il},0,\ldots,0)$. Thus there will be a unique vector $\overline{\omega}_i$ representing any $x_i \in [0,\beta_i]$. The choice of the integers k_i is discussed below. Also let

$$\Delta q_{ij} = q_i(jh_i) - q_i((j-1)h_i) \quad i=1,\ldots,n, \; j=1,\ldots,k_i \tag{10}$$

It follows that the linear function

$$\Gamma_i(x_i) = \sum_{j=1}^{k_i} \Delta q_{ij}\omega_{ij} \tag{11}$$

interpolates $q_i(x_i)$ at the points $x_i = jh_i$, $j = 0,1,\ldots,k_i$, and since $q_i(x_i)$ is concave it satisfies $\Gamma_i(x_i) \le q_i(x_i)$ for $x_i \in [0,\beta_i]$, where the vectors $\overline{\omega}_i$ are determined by (9). That is $\Gamma_i(x_i)$ is a piecewise linear underestimating function for $q_i(x_i)$.

We can therefore approximate the objective function $\Psi(x,y)$ in (SP) by the underestimating function

$$\sum_{i=1}^{n} \Gamma_i(x_i) + d^T y = \sum_{i=1}^{n}\sum_{j=1}^{k_i} \Delta q_{ij}\omega_{ij} + d^T y. \tag{12}$$

The global minimum problem (SP) can now be approximated by the following linear, zero-one, mixed integer problem in the continuous variables ω_{ij} and the zero-one variables z_{ij}:

$$\min \sum_{i=1}^{n}\sum_{j=1}^{k_i} \Delta q_{ij}\omega_{ij} + d^T y \tag{MI}$$

subject to

$$\sum_{i=1}^{n} h_i a_i \sum_{j=1}^{k_i} w_{ij} + A_2 y = b$$

$$0 \le \omega_{ij} \le 1, \; y \ge 0$$

$$\omega_{i,j+1} \le z_{ij} \le \omega_{ij}, \; j=1,\ldots,k_i-1, \; i=1,\ldots,n$$

$$z_{ij} \in \{0, 1\}$$

where a_i is the ith column of A_1.

We now show how to choose the k_i so that the relative error in the piecewise linear approximation is bounded by ε. Without loss of generality we may assume that

$$0 < \rho_n \leq \rho_{n-1} \leq \cdots \leq \rho_1 = 1 \tag{13}$$

and so we want to choose integers k_i such that

$$1 \leq k_n \leq k_{n-1} \leq \cdots \leq k_1 \tag{14}$$

Specifically choose k_i to be the smallest integer such that

$$k_i \geq \left(\frac{n}{\alpha}\rho_i\right)^{\frac{1}{2}} \tag{15}$$

where $\alpha = \varepsilon \sum_{i=1}^{n} \rho_i (1 + \eta_i)^2$, for the given tolerance ε.

It follows immediately that

$$\sum_{i=1}^{n} \frac{\rho_i}{k_i^2} \leq \alpha = \varepsilon \sum_{i=1}^{n} \rho_i (1 + \eta_i)^2 \tag{16}$$

The function $\Gamma_i(x_i)$, given by (11) interpolates $q_i(x_i)$ at the points $x_i = jh_i$, j=0,1, . . . , k_i. Since $q_i(x_i)$ is a concave, quadratic function it is easy to show that the error in this piecewise linear approximation satisfies

$$0 \leq q_i(x_i) - \Gamma_i(x_i) \leq \frac{1}{8}\lambda_i\left(\frac{\beta_i}{k_i}\right)^2,\ i=1,2,\ldots,n \tag{17}$$

Let $(\hat{x},\hat{y}) \in \Omega$ be the solution to

$$\min \sum_{i=1}^{n} \Gamma_i(x_i) + d^T y$$
$$(x,y) \in \Omega$$

Then $\sum_{i=1}^{n} \Gamma_i(\hat{x}_i) + d^T \hat{y} \le \Psi^* \le \phi(\hat{x}) + d^T \hat{y} = \Psi(\hat{x},\hat{y})$

and $0 \le \Psi(\hat{x},\hat{y}) - \Psi^* \le E(\hat{x})$ where $E(\hat{x}) = \phi(\hat{x}) - \sum_{i=1}^{n} \Gamma_i(\hat{x}_i) \ge 0.$

Theorem 8.4.1:

$$\frac{\Psi(\hat{x},\hat{y}) - \Psi^*}{\Delta\phi_m} \le \varepsilon \tag{18}$$

Proof: We have that

$$E(\hat{x}) = \phi(\hat{x}) - \sum_{i=1}^{n} \Gamma_i(\hat{x}_i) = \sum_{i=1}^{n} [q_i(\hat{x}_i) - \Gamma_i(\hat{x}_i)]$$
$$\le \frac{1}{8} \sum_{i=1}^{n} \lambda_i \frac{\beta_i^2}{k_i^2} \le \frac{1}{8} \lambda_1 \beta_1^2 \sum_{i=1}^{n} \frac{\rho_i}{k_i^2}$$

We also have from the previous lemma that

$$\Delta\phi_m \ge \frac{1}{8} \lambda_1 \beta_1^2 \sum_{i=1}^{n} \rho_i (1 + \eta_i)^2$$

Therefore

$$\frac{E(\hat{x})}{\Delta\phi_m} \le \left(\sum_{i=1}^{n} \rho_i (1 + \eta_i)^2\right)^{-1} . \sum_{i=1}^{n} \frac{\rho_i}{k_i^2} \le \varepsilon \quad \text{(from 16)}$$

and

$$\frac{\Psi(\hat{x},\hat{y}) - \Psi^*}{\Delta\phi_m} \le \frac{E(\hat{x})}{\Delta\phi_m} \le \varepsilon.$$

This theorem guarantees that the solution to the zero-one mixed integer problem (MI) will give an ϵ-approximate solution to the original global minimization problem (P).

8.5 Computational Algorithm and Analysis

The theoretical results described previously serve as the basis for an efficient computational algorithm for computing an ϵ-approximate solution to the problem (P). The algorithm first constructs and solves the linear underestimating approximation based on the enclosing rectangle. With good luck this will give an ϵ-approximate solution and we are done. If so, the global minimization problem is solved in the time required to solve a multiple-cost-row linear program with $(2n + 1)$ cost rows, m constraints and $n + k$ variables. If not, then the piecewise linear underestimating function is constructed and the corresponding 0-1 mixed integer linear program is solved. The solution to this problem is guaranteed to give an ϵ-approximate solution.

A simplified form of the computational algorithm is as follows:

1. Compute the eigenvalues $\lambda_i(Q)$ and the corresponding orthonormal eigenvectors u_i, $i=1, \ldots, n$.

2. Solve the multiple-cost-row LP with 2n cost rows. Evaluate $\Psi(x,y)$ at every vertex encountered.

3. Choose min $\Psi(x_i,y_i)$ as the incumbent function value (IFV). Construct $\Gamma(x)$. Compute $\Delta\phi_m$, ρ_i, $\overline{x}_i$.

4. Solve (LU) and get the vertex $(\hat{x},\hat{y})$. If $\Psi(\hat{x},\hat{y}) <$ IFV, set IFV:= $\Psi(\hat{x},\hat{y})$.

5. If IFV$-\Gamma(\hat{x})-d^T\hat{y} \leq \epsilon\Delta\phi_m$; stop. Incumbent is an ϵ-approximate solution.

6. Otherwise, construct the piecewise linear functions $\gamma_i(x_i)$, corresponding to given

tolerance ε. Solve (MI) to get an ε-approximate solution, using the incumbent to accelerate pruning.

It should be noted that the time required by step 1 of the algorithm is relatively insignificant, since all the eigenvalues and eigenvectors of a symmetric matrix with n = 100 can be computed in approximately 2 seconds of CPU time using EISPAK [SMIT76] on, for example, a Cyber 74/730.

In the process of solving the multiple-cost-row problem (MCR) the number of iterations required is $0(m^2)$. Each iteration gives a basis and corresponding vertex of Ω. Each such vertex is a candidate for the global minimum, and therefore $\Psi(x,y)$ is computed at each of these vertices. That vertex with the minimum Ψ is the incumbent and gives the IFV.

The availability of efficient packages for solving the 0-1 mixed integer linear program is an important factor in the practical implementation of this algorithm. For small problems ($n \leq 25$) the LINDO package [SCHR84] can be used, while for larger problems impressive reports on the package PIPX have been given [CROW82]. Substantial testing of this algorithm using these codes is needed in order to determine the average behavior of these 0-1 mixed integer codes for this class of problems.

The formulation of problem (P) as a mixed zero-one integer program (MI) requires a total of

$$N = \sum_{i=1}^{n} k_i - n \tag{19}$$

0-1, integer variables.

If all possible combinations of the integer variables were allowed, this problem would have a maximum of 2^N possibilities to be evaluated. However the structure of the problem is such that these 0-1 variables must satisfy

$$\omega_{i,j+1} \leq z_{ij} \leq \omega_{ij}$$

and therefore

$$z_{i,j+1} \le z_{ij} \quad j=1,\ldots,k_i-1$$

and thus the only possibilities for z_{ij} are

$$z_{ij} = 1,\ j=1,\ldots,l,\ z_{ij} = 0,\ j=l+1,\ldots,k_i-1 \text{ for some } l,\ 1 \le l \le k_i-1 \quad (20)$$

More explicitly any $\overline{z_i} = (z_{i1},\ldots,z_{ik_i-1}) = (1,1,\ldots,1,0,\ldots,0)$.
This requires also that $\overline{w_i} = (\omega_{i1},\ldots,\omega_{ik_i}) = (1,1,\ldots1,\omega,0,\ldots0)$ where $0 \le \omega \le 1$.

Because of the requirements (20), the maximum number of possible combinations is given by

$$\overline{N} = \prod_{i=1}^{n} k_i \le k_1^n$$

Note that $k_1^n = 2^r$ where $r = n\log_2 k_1$ and $2^r \ll 2^N$. Furthermore a good initial incumbent vector is available to start the integer program. This vector is obtained initially by the (MCR) and (LU) problems. This will give a good initial upper bound for the integer problem, which should result in the rapid pruning of many branches and therefore a relatively fast solution of the integer program.

Remark: By the choice of $k_i = k_i(\varepsilon)$ we have

$$\begin{aligned}
N &= \sum_{i=1}^{n} k_i - n \le \sum_{i=1}^{n} \left(\left(\frac{n}{\alpha}\rho_i\right)^{\frac{1}{2}} + 1\right) - n \\
&= \left(\frac{n}{\alpha}\right)^{\frac{1}{2}} \sum_{i=1}^{n} \rho_i^{\frac{1}{2}} \\
&\le \left(\frac{n}{\alpha}\right)^{\frac{1}{2}} n^{\frac{1}{2}} \left(\sum_{i=1}^{n} \rho_i\right)^{\frac{1}{2}} \quad \text{by the Cauchy-Schwarz inequality} \\
&= n\left(\sum_{i=1}^{n} \rho_i/\alpha\right)^{\frac{1}{2}} \\
&= n\left(\frac{1}{\varepsilon}\,\sigma(\rho,\eta)\right)^{\frac{1}{2}} \quad \text{(see eqs (7) and (15))}
\end{aligned}$$

Hence a worst case upper bound for the number N of 0-1 integer variables is given by

$$N \le n\left(\frac{\sigma(\rho,\eta)}{\varepsilon}\right)^{\frac{1}{2}}$$

for a given tolerance ε. Note that $\sigma(\rho,\eta) \in [\frac{1}{4},1]$

Example 8.5.1 :

Consider the 3-dimensional problem

$$\min_x \phi(x) = \sum_{i=1}^{3} q_i(x_i)$$

s.t.

$$x \in \Omega = \{x \in R^3 \colon \; Ax \le b, \; x \ge 0\}$$

with

$$A = \begin{bmatrix} 10.0 & 0.2 & -0.1 \\ -0.3 & 9.0 & 0.2 \\ -0.1 & 0.4 & 11.0 \\ 6.0 & 8.0 & 9.0 \end{bmatrix}, \quad b^T = (11,8,12,18)$$

$q_i(x_i) = c_i x_i - \frac{1}{2}\lambda_i x_i^2$,$c_1 = 1.25$, $c_2 = 2.5$ $c_3 = 5.0$, $\lambda_1 = 5$, $\lambda_2 = 10$, $\lambda_3 = 15$.

First by solving the (MCR) problems $\min_{1 \le i \le 3} x_i$, $\max_{1 \le i \le 3} x_i$, $x \in \Omega$ we find the smallest rectangle containing Ω to be $R_x = \{x \colon 0 \le x_1 \le 1.111, \; 0 \le x_2 \le 0.9249, \; 0 \le x_3 \le 1.101\}$

The linear underestimating function on the rectangle R is given by $\Gamma(x) = -(1.5275x_1 + 2.1245x_3 + 3.2585x_3)$.

Then, $\min_{x \in \Omega} \Gamma(x) = -5.685$ at $v_1 = (0.2434, 0.8734, 1.06136)$ and $\phi(v_1) = -4.6163$.

Divide each of the intervals $[0,\beta_i]$ into two subintervals (i=1,2,3), in order to obtain an improved approximate underestimating function.

This new approximate piecewise linear function is formulated as a zero-one mixed integer linear program. Since k_i=2 for i=1,2,3 we introduce 3 zero-one variables. By solving the corresponding integer program we obtain $\min \Gamma(w,z) = -5.2835$ at $v^* = (1.111,0,1.101)$. Note that $\phi(v^*) = -5.2835$ and so the optimal solution has been found.

8.6 Exercises

1. Find the global optimum of the following problem:

$$\min f(x) = x - x^2 - 4y$$

$$\text{s.t. } x+y \le 10, \; -x+2y \le 5$$

$$x+4y \ge 13, \; x,y \ge 0.$$

2. Consider the problem

$$\text{global} \min_{(x,y)\in P} F(x,y) = \phi(x) + d^T y$$

where $\phi(x)$ is a quadratic concave function defined on R^n, $d \in R^m$ and P is a bounded polyhedron in R^{n+m}.
Let (x_l, y_l) be a vertex that solves the linear program

$$\min_{(x,y)\in P} d^T y.$$

Can you find conditions which insure that (x_l, y_l) is also a global minimum of $F(x,y)$ over P?

3. Let $f(x)$ be a concave function defined on a bounded polyhedron P in R^n such that $|f(x) - f(y)| \le M \, \|x - y\|$ for all $x,y \in P$. Consider a set of vertices $v_1, \ldots, v_k$ and let

$$m = f(v_0) = \min \{f(v_1), \ldots, f(v_k)\}.$$

Denote by S_i the spheres with center v_i and radius $r_i = (f(v_i) - m)/M$. Prove that if $P \subset \bigcup_{i=1}^{k} S_i$ then v_0 is the global minimum of $f(x)$ over P.

8.7 References

[CROW82] Crowder, H., Johnson, E.L., and Padberg, M.W. *Solving large-scale zero-one linear programming problems*. Oper. Res. Vol.31, No.5 (1982), 803-834.

[FALK76] Falk, J.E., and Hoffman, K.R. *A successive underestimating method for concave minimization problems*. Math. Oper. Res. 1 (1976), 251-259.

[KALA84] Kalantari, B. *Large scale concave quadratic minimization and extensions*. PhD thesis, Computer Sc. Dept., Univ. of Minnesota, March 1984.

[PARD85] Pardalos, P.M. *Integer and separable programming techniques for large scale global optimization problems*. PhD thesis, Computer Sc. Dept., Univ. of Minnesota, July 1985.

[PARD86] Pardalos, P.M. and Rosen, J.B. *Methods for global concave minimization: A bibliographic survey*. SIAM Review 28 (1986), 367-379.

[ROSE83] Rosen, J.B. *Global minimization of a linearly constrained concave function by partition of feasible domain*. Math. Oper. Res. 8 (1983), 215-230.

[ROSE84] Rosen, J.B. *Performance of approximate algorithms for global minimization*. Math. Progr. Study 22 (1984), 231-236.

[ROSE85] Rosen, J.B. *Computational solution of large scale constrained global minimization problems*. Numerical Optimization 1984. (P. T. Boggs, R. H. Byrd, R. B.Schnabel, Eds) SIAM, Phila. PA (1985), 263-271.

[ROSE86] Rosen, J.B. and Pardalos, P.M. *Global minimization of large scale constrained concave quadratic problems by separable programming*. Math. Progr. 34 (1986), 163-174.

[SCHR84] Schrage, L. *Linear Integer and Quadratic Programming with LINDO*. Scientific Press, Palo Alto, CA (1984).

[SMIT76] Smith, B.T., Boyle, J., Garbow, B., Ikebe,Y., Klema, V., and Moler, C. *Matrix Eigensystem Routines-EISPACK Guide*. Lecture Notes in Computer Sc., Vol. 6, Springer-Verlag, N. Y. (1976).

[ZILV83] Zilverberg, N. *Global minimization for large scale linearly constrained systems*. PhD thesis, Computer Sc. Dept. Univ. of Minnesota, Dec. 1983.

Chapter 9 Global Minimization of Indefinite Quadratic Problems

9.1 Introduction

In this chapter we are going to consider several aspects of indefinite quadratic programming problems. The general problem is of the form

$$\text{global} \min_{x \in \Omega} \phi(x) = c^T x + \frac{1}{2} x^T Q x \qquad \text{(IQP)}$$

where $\Omega = \{x : Ax \leq b,\ x \geq 0\} \subseteq R^n$ is a bounded polyhedral set, $Q_{n \times n}$ is an indefinite symmetric matrix and $A \in R^{m \times n}$, $b \in R^m$, and $x, c \in R^n$.

Quadratic programming has been a very old and important problem of mathematical programming. It has numerous applications in many diverse fields of science and technology, and plays a key role in many nonlinear programming methods.

Apart from its importance as a mathematical programming problem, recently it has found applications in VLSI chip design. Certain aspects of physical chip design can be formulated as indefinite quadratic problems. For example the compaction problem can be stated as the global minimization of an indefinite quadratic function, that is: *Find the minimum chip area, subject to constraints (linear and nonlinear).* The constraints come from geometric design rules, from distance and connectivity requirements between various components of the circuit, and possibly from user specified constraints. Regarding these applications see [CIEL82], [KEDE83], [MALI82], [SOUK84] and [WATA84]. For a general survey of optimization techniques used in integrated circuit design see [BRAY81]. However, because of the computational difficulty of the problem, so far only heuristics and some approximate alternatives have been developed for the problems arising from the above applications.

Another class of problems closely related to the indefinite quadratic problem is the linear complementarity problem. Lately some new approaches using global optimization techniques have been developed (see for example [PARD85A], [THOA83]).

Traditional nonlinear programming methods usually obtain local solutions when applied to (IQP). In many applications the global optimum or a good approximation to the global optimum is required. Recently some new approaches have been developed for finding the global optimum of indefinite quadratic programs. One such approach is given in [KALA84].

Kough [KOUG79] proposed an algorithm for the indefinite quadratic problem that uses a generalized Benders cut procedure which was developed by Geoffrion [GEOF72]. The problem considered is in the following (polar) form:

$$\max \ x^T x - y^T y \qquad \text{(PF)}$$

$$s.t. \quad Ax+By+c \geq 0, \ x \in R^n, y \in R^t$$

and A, B, are $m \times n$ and $m \times t$ matrices respectively and $c \in R^m$. The separability of the objective function into x, y variables (convex and concave part), suggests a natural decomposition for the Benders cut problem. If x_0 is a fixed feasible point we define the following convex problem $P(x_0)$:

$$\max_y -y^T y$$

$$\text{s.t.} \quad By \geq -(Ax_0+c)$$

Let y_0 be the optimal solution to the above problem. Using problem $P(x_0)$ we define the function

$$v(x_0) = (x_0)^T x_0 - (y_0)^T y_0$$

Then problem (PF) is equivalent to the next problem $P_v(x)$:

$$\max_x v(x)$$

$$\text{s.t.} \quad x \in R = \{x : Ax+By+c \geq 0 \text{ for some } y\}$$

Here R is the projection of $\{(x,y) : Ax+By+c \geq 0\}$ into the x-space. When the Benders cut method is applied to (PF) it generates approximations $v_k(x)$ of $v(x)$ and R_k of R satisfying $v_{k_1}(x) \geq v_{k_2}(x) \geq v(x)$ for all x and $R_{k_1} \supseteq R_{k_2} \supseteq R$ for $k_2 \geq k_1$. At the kth step the approximate problem

$$\max v_k(x)$$

$$\text{s.t.} \ x \in R_k$$

is solved to obtain an approximate optimal x_k. If $x_k \in R$, a Benders feasibility cut is generated in order to obtain R_{k+1}. Otherwise a Benders cut is generated to obtain $v_{k+1}(x)$. This algorithm will give ε-finite convergence [GEOF72]. Furthermore Kough developed exact cuts and proposed a modified finite algorithm.

Tuy [TUY84] uses a method based on Bender's decomposition technique for the global minimization of the difference of two convex functions.

Cirina [CIRI86] proposed an algorithm for the indefinite quadratic program (IQP) that in a finite number of iterations converges to a global minimum. The proposed algorithm produces a decreasing sequence of Kuhn-Tucker points by solving a sequence of associated linear complementarity problems (see chapter 3) with data

$$M = \begin{bmatrix} 2(Q+Q^T) & A^T & 0 \\ -A & 0 & 0 \\ -c & b & 0 \end{bmatrix}, \quad q^T = (c, b, 2\phi(x) - \varepsilon)$$

each with fixed x and ε. Details on the properties of the algorithm and preliminary computational results can be found in [CIRI86].

In this chapter, we use techniques from global concave minimization to obtain good approximate solutions for indefinite quadratic programming problems. We show how the problem (IQP) can be reduced to a separable quadratic problem by solving a multiple cost-row linear program, where the cost rows are the orthogonal eigenvectors of Q. The equivalent transformed problem has an objective function that is the sum of concave, convex and linear functions.

A linear underestimating function of the concave part is then easily obtained, and a separable convex problem is solved to obtain $\hat{x}$, a candidate for the global minimum. Lower and upper bounds on the global optimum are obtained. We also derive bounds for the relative error using appropriate scaling factors.

Next, we consider branch and bound techniques that improve the lower and upper bounds for the global optimum, and eliminate parts of the feasible domain from further consideration.

When the solution obtained so far is not a satisfactory approximation to the global minimum, then we use piecewise linear approximations and the original problem is reduced to a linear zero-one mixed integer program. The number of zero-one integer variables used depends only on the concave part of the objective function.

In section 9.6 a number of computational results are reported using the Cray 1S supercomputer. Finally, in the last section the more general large-scale indefinite quadratic programming problem is introduced.

9.2 Partition of the objective function

We can assume that Q is a symmetric matrix. Therefore all eigenvalues $\lambda_1, \ldots, \lambda_n$ of Q are real numbers. Without loss of generality suppose that

$$\lambda_1, \ldots, \lambda_l \neq 0$$

$$\lambda_{l+1} = \cdots = \lambda_n = 0$$

We formulate now an equivalent problem where the objective function is separable. We do that by first solving a multiple cost-row linear program of the form

$$\min_{x \in \Omega} a^T x \qquad \text{(MCR)}$$

with $a = u_i, -u_i, i=1, \ldots, n$, where u_i are the orthogonal eigenvectors of Q.

Let $\max_{x \in \Omega} u^T{}_i x = \overline{\alpha}_i$ and $\min_{x \in \Omega} u^T{}_i x = \underline{\alpha}_i$ for $i=1, \ldots, n$. If $U=[u_1, \ldots, u_n]$ is the orthogonal matrix of eigenvectors of Q, then $Q=UDU^T$ where $D=diag(\lambda_1, \ldots, \lambda_n)$, and so we can easily construct an affine transformation to obtain an equivalent problem in separable form. Other transformations are possible that can put the (IQP) in separable form. However here we preserve the eigenstructure of the quadratic form which is essential for the error analysis of the proposed approximations.

This transformation is given by $x \leftarrow U^T x - \underline{\alpha}$ (or $x_i \leftarrow u_i{}^T x - \underline{\alpha}_i$). Also a rectangle $R=\{x: 0 \leq x_i \leq \beta_i, i=1, \ldots, n\}$ of minimum volume that contains P, is obtained using (MCR). Here $\beta_i = \overline{\alpha}_i - \underline{\alpha}_i$, $i=1,\ldots,n$.

From now on, for simplicity of notation, we consider the following separable indefinite quadratic problem:

$$\text{global} \min_{x \in \Omega} \phi(x) = \phi_1(x) + \phi_2(x) \qquad \text{(ISP)}$$

where $\Omega = \{x: Ax \le b, 0 \le x_i \le \beta_i, 1=1, \ldots, n\} \subseteq R^n$.

The original problem has been transformed into a separable form where the objective function is partitioned to $\phi(x)=\phi_1(x)+\phi_2(x)$ with

$\phi_1(x) = \sum_{i=1}^{k} \theta_i(x_i) = \sum_{i=1}^{k} (c_i x_i - \frac{1}{2}\lambda_i x_i^2)$, the concave part ($\lambda_i > 0$, $i=1, \ldots, k$),

and

$\phi_2(x) = \sum_{i=k+1}^{n} \theta_i(x_i) = \sum_{i=k+1}^{n} (c_i x_i + \frac{1}{2}\lambda_i x_i^2)$, the convex part ($\lambda_i \ge 0$, $i=k+1, \ldots, n$).

(Note that $\sum_{i=l+1}^{n} c_i x_i$, is the linear part corresponding to zero eigenvalues.)

Since Q is an indefinite symmetric matrix we have at least one negative and one positive eigenvalue. Therefore the partition of the objective function into a concave and a convex part it is always valid.

Problem (ISP) has an objective function that can be written as a difference of two convex functions. From the computational complexity point of view this is an NP-hard problem and can be reduced to a concave minimization problem. The difficulty with (ISP) is that we may have many local minima. In fact we can construct nontrivial problems of this form where the number of local minima is 2^m, if m is the number of negative eigenvalues. For such a class of problems see [CIRI85].

9.3 Approximate Methods and Error Bounds

We consider next the development of an algorithm based on concave programming. The main idea is to obtain approximations of the concave part of the objective function and use them to solve the problem efficiently.

For each concave function $\theta_i(x_i) = c_i x_i - \frac{1}{2}\lambda_i x_i^2$, $i=1, \ldots, k$ the linear functions $\gamma_i(x_i) = (c_i - \frac{1}{2}\lambda_i \beta_i)x_i$, $i=1, \ldots, k$ satisfy $\gamma_i(x_i) \le \theta_i(x_i)$ on Ω. We also have $\gamma_i(0) = \theta_i(0)$ and $\gamma_i(\beta_i) = \theta_i(\beta_i)$ for $i=1, \ldots, k$. Therefore the function

$f(x) = \sum_{i=1}^{k} \gamma_i(x_i) + \phi_2(x)$ is a convex underestimating function to the original objective function $\phi(x)$. Furthermore we have the following error bound:

Theorem 9.3.1. $|\phi(x) - f(x)| \leq \frac{1}{8} \sum_{i=1}^{k} \lambda_i \beta_i^2$

Proof: It is easy to verify that the following is true:

$$|\theta_i(x_i) - \gamma_i(x_i)| \leq \frac{1}{2} \lambda_i |x_i(x_i - \beta_i)|.$$

Then, the maximum of $|x_i(x_i - \beta_i)|$ for $0 in \leq in x_i \leq \beta_i$, occurs at the midpoint $x_i = \beta_i/2$. Therefore we have

$$|\theta_i(x_i) - \gamma_i(x_i)| \leq \frac{1}{2} \lambda_i (\frac{\beta_i}{2})^2 = \frac{1}{8} \lambda_i \beta_i^2.$$

Note that the bound in this theorem is the best possible.

This error bound says that the approximation depends on the sum of the eigenvalues corresponding to the concave part of the objective function and the sides of the rectangle $R = \{x : 0 \leq x_i \leq \beta_i,\ i = 1, \ldots, k\}$. In other words the approximation depends on the curvature of the concave functions and the geometry of the feasible domain.

Next we solve the following convex separable problem:

$$\min_{x \in \Omega} f(x) = \sum_{i=1}^{k} \gamma_i(x_i) + \phi_2(x) \qquad \text{(CP)}$$

This problem can be solved efficiently using the MINOS 5.0 code [MURT83]. Good starting points are available from the solution of (MCR) problem. Since (CP) is a convex problem the solution is going to be global. Let $\hat{x}$ be the solution obtained by solving (CP). Then if ϕ^* is the global optimum to the original problem we have

$$0 \leq \phi(\hat{x}) - \phi^* \leq \phi(\hat{x}) - (\hat{x}) \leq \frac{1}{8} \sum_{i=1}^{k} \lambda_i \beta_i^2$$

Also we have that

$$f(\hat{x}) \leq \phi^* \leq \phi(\hat{x})$$

If $\phi(\hat{x}) - f(\hat{x})$ is small enough we accept $\hat{x}$ as an approximate solution.

Observe that $\hat{x}$ is a global solution to the approximate convex subproblem. In case that $\phi(\hat{x}) - f(\hat{x})$ is not small enough we apply MINOS to the original problem (ISP) with $\hat{x}$ as a starting point to obtain a local minimum solution. The new solution can be a better one. Although this is a heuristic step it seems to work in a large class of problems.

Since the error in the above approximation depends only on the concave part ϕ_1, we can obtain an appropriate scaling factor $\Delta\phi_1$, which measures the range of ϕ_1 over the rectangle R_1, where $R_1 = \{x : 0 \leq x_i \leq \beta_i, i=1, \ldots, k\}$ (see also [PARD85] and [ROSE86]). Then an a-priori bound on the relative error in $\phi(\hat{x})$ is given by the following theorem:

Theorem 9.3.2: $\dfrac{\phi(\hat{x}) - \phi^*}{\Delta\phi_1} \leq \sigma$ where $0.25 \leq \sigma in \leq 1$.

Proof: (see chapter 8)

9.4 Branch and Bound Bisection Techniques

We continue by using some new branch and bound techniques for (IQP). Branching refers to the partition of the feasible domain by bisection in certain directions, and bounding refers to the determination of lower and upper bounds for the global optimum.

The main idea here is to bisect the rectangle R that contains the feasible domain Ω, along each axis corresponding to concave variables. In this way we can obtain better approximations of the concave part of the objective function and we can eliminate part of the feasible domain from further consideration. The bisection procedure is done only until we reach a certain number of bisections, or no further elimination is possible. Preliminary computational results indicate that this approach gives very good lower and upper bounds for the global solution.

Elimination Procedure

Let $l \in \{1, \ldots, k\}$ and let β_{l3} be the midpoint of the interval $[\beta_{l1}, \beta_{l2}]$. (Initially $\beta_{i1}=0$ and $\beta_{i2}=\beta_i$, $i=1, \ldots, k$.) Define γ_{l1} to be the line through the points $(\beta_{l1}, \theta_l(\beta_{l1}))$ and $(\beta_{l3}, \theta_l(\beta_{l3}))$. Similarly we define the line γ_{l2} through $(\beta_{l3}, \theta_l(\beta_{l3}))$ and $(\beta_{l2}, \theta_l(\beta_{l2}))$. Consider now the convex functions

$$f_{l1}(x) = \sum_{i \neq l} \gamma_i(x_i) + \gamma_{l1}(x_l) + \phi_2(x)$$

$$f_{l2}(x) = \sum_{i \neq l} \gamma_i(x_i) + \gamma_{l2}(x_l) + \phi_2(x)$$

Solve now the following convex programs ($2k$ objective functions):

$$\min_{x \in \Omega} f_{li}(x), \quad l=1, \ldots, k, \quad i=1,2$$

to obtain solutions $\hat{x}_{li}$. Denote $f_{li}(\hat{x}_{li})$ by $\hat{f}_{li}$.

Theorem 9.4.1. If $\hat{f}_{l1} > \phi(\hat{x}_{l1})$ then we can eliminate $[\beta_{l1}, \beta_{l3}]$ from further consideration. Similarly if $\hat{f}_{l2} > \phi(\hat{x}_{l2})$ then we can eliminate $[\beta_{l3}, \beta_{l2}]$.

Proof: (If $x \in R^n$ denote by $(x)_i$ the i-component of x). First note that if $\hat{f}_{l1} \geq \phi(\hat{x}_{l1})$ then $(\hat{x}_{l1})_l \in [\beta_{l3}, \beta_{l2}]$ since $\gamma_{l1}(x_l) \leq \theta_l(x_l)$ in $[\beta_{l1}, \beta_{l3}]$. Suppose x^* is the global minimum and that $(x^*)_l \in [\beta_{l1}, \beta_{l3}]$. Then,

$f_{l1}(x^*) \leq \phi^*$ since $f_{l1} \leq \phi$ on $[\beta_{l1}, \beta_{l3}] \cap \Omega$,

$\phi^* \leq \phi(\hat{x}_{l1})$ since ϕ^* is the global optimum, and

$\phi(\hat{x}_{l1}) \leq f_{l1}(\hat{x}_{l1})$ by assumption.

Therefore, $f_{l1}(x^*) < f_{l1}(\hat{x}_{l1})$ which is impossible since $\hat{x}_{l1}$ is the minimum. Hence $(x^*)_l$ does not belong to $[\beta_{l1}, \beta_{l3}]$. The proof is similar for f_{l2}.

Error Bounds

Let $\hat{f} = \max_{l} \{\min_{i=1,2} \hat{f}_{li}\}$, $\hat{\phi} = \min_{li} \phi(\hat{x}_{li})$ for $l=1, \ldots, k$.

Theorem 9.4.2. $\hat{f} \le \phi^* \le \hat{\phi}$

Proof: The upper bound is obvious. Consider now the lower bound. Let $\Omega = \Omega_{l1} \cup \Omega_{l2}$ be a partition of the feasible domain. We have

$$\hat{f}_{li} \le \min_{x \in \Omega_{li}} f_{li}(x) \le \phi^* \quad ,i=1, \text{ or } 2.$$

Therefore $\min_{i=1,2} \hat{f}_{li} \le \phi^*$ for $l=1, \ldots, k$, so that $\hat{f} \le \phi^*$.

Suppose the elimination occurs at j intervals $(0 \le j \le k)$, with corresponding variables $x_{i1}, x_{i2}, \ldots, x_{ij}$. Then we define the convex function

$$f_1(x) = \sum_{i \notin E} \gamma_i(x_i) + \sum_{l \in E} \gamma_l(x_l) + \phi_2(x)$$

where $E = \{i1, \ldots, ij\} \subseteq \{1, \ldots, k\}$ and $\gamma_l = \gamma_{l1}$ or γ_{l2} depending on the elimination of $[\beta_{l3}, \beta_{l2}]$ or $[\beta_{l1}, \beta_{l3}]$. Solve the following convex program

$$\min_{x \in \Omega} f_1(x)$$

to obtain the point $\bar{x}$. Note that $f_{l1}(x) \le f_1(x)$ in the case where that the interval $[\beta_{l3}, \beta_{l2}]$ is eliminated. Similarly for the other convex functions f_{li}.

Theorem 9.4.3. Suppose $(\bar{x})_l \in [\beta_{l3}, \beta_{l2}]$ (i.e. an eliminated interval). Then $[\beta_{l3}, \beta_{l2}]$ cannot contain a point x with $\phi(x) \le \hat{\phi}$.

Proof: Call $\Omega_l = \Omega \cap [\beta_{l3}, \beta_{l2}]$. Since $(\bar{x})_l \in [\beta_{l3}, \beta_{l2}]$ we have that

$$\min_{x \in \Omega_l} f_1(x) \le \min_{x \in \Omega} f_1(x) = f_1(\bar{x}) \quad \text{and } f_1(\bar{x}) \ge f_{l1}(\bar{x}) \ge \min_{x \in \Omega} f_{l1}(x) \ge \hat{\phi}.$$

The proof is now obvious.

Here we have an improvement in the size of the error bound:

$$|f_1(x) - \phi(x)| \le \frac{1}{8} \sum_{x \notin E} \lambda_i (\beta_{i1} - \beta_{i2})^2 + \frac{1}{32} \sum_{i \in E} \lambda_i (\beta_{i1} - \beta_{i2})^2$$

We can continue the elimination procedure by updating the lower and upper bounds of the rectangular domain:

If $f_{l1}(\hat{x}_{l1}) \geq \phi(\hat{x}_{l1})$ then $\beta_{l1} \leftarrow \beta_{l3}$.

If $f_{l2}(\hat{x}_{l2}) \geq \phi(\hat{x}_{l2})$ then $\beta_{l2} \leftarrow \beta_{l3}$.

If we can no longer continue with the elimination of parts of the feasible domain then we use the MINOS program, using as starting point the one that gives the smallest upper bound, to obtain a better solution $\hat{x}$.

9.5 Piecewise Linear Approximation

If $\hat{x}$ is not a satisfactory solution, we can always obtain better approximations by increasing the computational effort.

We use a piecewise linear approximation for each function θ_i. For $[0,\beta_i]$ let $p_i(x_i)$ be the piecewise linear function that interpolates the corresponding quadratic function at $0, \beta_i/k_i, 2\beta_i/k_i, \ldots, \beta_i$ with k_i some positive integer (as a first step we can use $k_i=2$). Let $p(x) = \sum_{i=1}^{l} p_i(x_i) + \sum_{i=l+1}^{n} c_i x_i$ be the piecewise linear approximation. Then the following is true:

Theorem 9.5.1.

$$|\phi(x) - p(x)| \leq \frac{1}{8}\sum_{i=1}^{l} \lambda_i \left(\frac{\beta_i}{k_i}\right)^2 .$$

Proof: Similar as in Theorem 9.3.1.

The piecewise linear approximation of the convex part can easily be formulated as a linear function ([MEYE83], [THAK78]).

For the piecewise linear approximation of the concave part we apply the method described in [ROSE86] to formulate this as a linear function by introducing zero-one variables. This can be done by introducing the new variables ω_{ij} and z_{ij} such that

$$x_i = \frac{\beta_i}{k_i}\sum_{j=1}^{k_i} \omega_{ij}, \quad i=1, \ldots, l$$

$$\omega_{i,j+1} \le z_{ij} \le \omega_{ij}, \; j=1,\dots,k_i-1, \; i=1,\dots,l$$

$$0 \le \omega_{ij} \le 1, \; z_{ij} \in \{0, 1\}$$

Then the piecewise linear function that interpolates the concave part $\phi_2(x)$ is reduced to the linear function

$$\Gamma(\omega) = \sum_{j=1}^{k_i} \Delta\theta_{ij}\omega_{ij}$$

where $\Delta\theta_{ij} = \theta_i(jh_i) - \theta_i((j-1)h_i)$, $i=1,\dots,l$, $j=1,\dots,k_i$.

The resulting approximate problem will be a zero-one mixed integer linear program. Note that we introduce integer variables only for the concave functions. For appropriately chosen integers $k_i = k_i(\varepsilon)$, we can guarantee that an ε-approximate solution to (QSP) will be obtained for any specified $\varepsilon > 0$.

9.6 Algorithm and Computational Results

Based on the above theoretical results, we now present the computational algorithm:

1. Compute the eigenvalues and eigenvectors of Q.

2. Solve the multiple-cost-row linear program $\min_{x \in \Omega} \pm u_i^T x$ to obtain the equivalent separable problem and the rectangle $R \supset \Omega$. Find the linear underestimator of the concave part. Solve the corresponding convex problem to get $\hat{x}$. If $\phi(\hat{x}) - f(\hat{x}) \le \varepsilon \Delta\phi_1$, stop.

3. Bisect the rectangle along each axis corresponding to concave variables. Obtain new linear approximations and solve the new convex problems. Repeat until a satisfactory solution is obtained, or no further bisection is possible.

4. Use piecewise linear approximation. Formulate and solve as a 0,1 mixed integer linear program. We introduce integer variables only for the concave variables.

Computational Results

The computational results described below were all obtained using the Cray 1S supercomputer (with front-end a Cyber 845). The Cray 1S is a single processor machine with one million 64-bit words of main memory and a vector processing capability of more than 100 megaflops.

Test Problems: All test problems were in separable form. The constraint matrix A is sparse (8 nonzero elements per column). All nonzero elements of A are positive integers in [1, 9]. The eigenvalues λ_i are specified or randomly generated integers in [0, 99]. Similarly the $\overline{x}_i \in [-99, 99]$. All test problems have the following form:

$$\text{global} \min_{x \in \Omega} -\frac{1}{2}\sum_{i=1}^{k} \lambda_i (x_i - \overline{x}_i)^2 + \frac{1}{2}\sum_{i=k+1}^{n} \lambda_i (x_i - \overline{x}_i)^2$$

$$\Omega = \{x : Ax \le b, x \ge 0\} \subseteq R^n$$

For the multiple-cost-row linear programs and the related convex separable programs we used MINOS [MURT83]. Initially we find the rectangle $R \supset \Omega$, by solving the multiple-cost-row linear program $\min_{x \in \Omega} \pm x_i$, $i=1, \ldots, n$. In our implementation of the algorithm we apply only steps 2 and 3. We solved 17 problems, the smallest of dimension 60 (10 concave and 50 convex variables) and the largest of dimension 330 (30 concave and 300 convex variables). The CPU time was 48.5 secs for n=60, to 1500 secs for the largest, n=330, problem. The relative error (in percentage) ranged from 0.015 to 4.9. These preliminary computational results indicate the practicality of our approach. Computational results are also reported in [PARD85A, B].

The next table summarizes the computational results.

k	l	m	CPU secs	% R. E.
10	50	30	54.6	1.46
10	50	20	56.6	1.81
10	60	20	154	.113
10	60	30	48.5	1.87
10	70	30	80.5	1.02
10	80	40	77.8	.015
10	100	50	89.5	1.45
20	40	20	97.5	.078
20	50	20	305.5	.22
20	60	30	208.8	.067
20	70	30	199	4.9
40	80	40	457	.52
20	150	30	325	4.05
20	150	50	436	1.03
30	150	50	1000	0.158
20	200	50	178	4.4
30	300	50	1500	2.8

Table 1

k= number of concave variables
l= number of convex variables
m= number of constraints
% R.E. = % Relative Error

9.7 Concluding Remarks

We note that in our approach the main computational effort required to obtain good approximations to the global solution of the constrained indefinite quadratic problem depends mainly on the concave part, that is, on the negative eigenvalues of the quadratic form.

Computationally we can easily implement the algorithm for the solution of the convex approximate subproblems and the branch and bound techniques. The MINOS (Version 5.0) large-scale optimization system can be used for the convex quadratic problems [MURT83]. Also, available efficient packages for solving 0-1 mixed integer linear programs can be used [CROW82]. For a small number of integer variables the piecewise approximation will be a computationally practical step. However when we have a large number of integer 0-1 variables we can only solve the relaxed linear program to obtain an approximate solution.

The techniques developed above can be extended for the solution of large scale indefinite quadratic problems of the form:

$$\text{global} \min_{(x,y)\in\Omega} F(x,y) = c^T x + \frac{1}{2} x^T Q x + d^T y$$

where $\Omega = \{(x,y)\colon Ax+By \le b,\ x \ge 0,\ y \ge 0\} \subseteq R^{n+k}$,is a bounded polyhedral set, $Q_{n\times n}$ is an indefinite symmetric matrix, $x \in R^n$ and $y \in R^k$.

Here k may be much larger than n. Since the approximations involve only the nonlinear part of the objective function, we can solve very large-scale indefinite quadratic problems of that form.

9.8 Exercises

1. Prove that the indefinite quadratic problem

$$\text{global} \min_{x\in P} c^T x + x^T Q x$$

where $Q_{n\times n}$ is a symmetric matrix and P is a bounded polyhedron in R^n, can be expressed as a jointly constrained bilinear program of the form:

$$\min x^T y$$

$$\text{s.t. } y - Qx = c,\ x \in P$$

2. Let P be a bounded polyhedron in R^n and let $l_1(x) = a^T x + \alpha$, $l_2(x) = b^T x + \beta$ be two linear functions satisfying $l_1(x) > 0$, $l_2(x) > 0$ for all $x \in P$. Prove that the next (indefinite) quadratic problem

$$\text{global} \min_{x \in P} f(x) = l_1(x).l_2(x)$$

has a solution that occurs at some vertex of P.

3. Solve the problem

$$\text{global min } f(x,y) = x^2 - y^2 + y$$

$$\text{s.t. } 1 \leq x+y \leq 2, \;\; x,y \geq 0.$$

9.9 References

[BENA85] Benacer, R. and Tao, P.T. *Global maximization of a nondefinite quadratic function over a convex polyhedron.* FERMAT Days 1985, Mathematics for Optimization (J.-B. Hirriart-Urruty, editor), Elsevier Sci. Publ., 65-77.

[BRAY81] Brayton, R.K., Hachtel, G.D., and Sangiovanni-Vincentelli, A.L. *A survey of Optimization Techniques for Integrated-Circuit Design.* Proceedings of the IEEE, Vol. 69, No. 10 (1981), 1334-1362.

[CIEL82] Ciesielski, M.J. and Kinnen, E. *An Analytic Method for Compacting Routing Area in Integrated Circuits.* Proceedings of the 19th Design Automation Conference, Las Vegas, NV. (1982), 30-37.

[CIRI85] Cirina, M. *A class of nonlinear programming test problems.* Working paper, Dipart. di Informatica, Torino (1985).

[CIRI86] Cirina, M. *A finite algorithm for global quadratic minimization.* Working paper, Dipart. di Informatica, Torino (1986).

[CROW82] Crowder, H., Johnson, E.L. and Padberg, M.W. *Solving large-scale zero-one linear programming problems.* Oper. Res. Vol. 31, No. 5 (1982), 803-834.

[GEOF72] Geoffrion, A. *Generalized Bender's Decompositions.* J. Optimiz. Theory Appl. 10 (1972), 237-260.

[KALA84] Kalantari, B. *Large scale concave quadratic minimization and extensions.* PhD thesis, Computer Sci. Dept., University of Minnesota 1984.

[KEDE83] Kedem, G. and Watanabe, H. *Optimization Techniques for IC Layout and Compaction.* Proceedings IEEE Intern. Conf. in Computer Design: VLSI in Computers (1983), 709-713.

[KOUG79] Kough, P.F. *The Indefinite Quadratic Programming Problem.* Oper. Res. Vol. 27, No.3 (1979), 516-533.

[MALI82] Maling, K., Mueller, S.H., and Heller, W.R. *On finding most optimal rectangular package plans.* Proceedings of the 19th Design Automation Conference, Las Vegas, NV. (1982), 663-670.

[MUEL79] Mueller, R.K. *A method for solving the indefinite quadratic programming problem.* Manag. Science Vo. 16, No. 5 (1979), 333-339.

[MEYE83] Meyer, R.R. *Computational Aspects of Two-segment Separable Programming.* Math. Progr. 26 (1983), 21-32.

[MURT83] Murtagh, B.A., and Saunders, M.A. *MINOS 5.0 User's Guide.* Tech. Rep. SOL 83-20 (1983), Dept. of Oper. Res., Stanford Univ.

[PARD86] Pardalos, P.M. and Rosen, J.B. *Methods for global concave minimization: A bibliographic survey.* SIAM Review 28 (1986), 367-379.

[PARD85A] Pardalos, P.M. and Rosen, J.B. *Global Optimization Approach to the Linear Complementarity Problem.* Tech. Report 84-37 (revised Aug. 1985) Computer Sci. Dept. Univ. of Minnesota.

[PARD85B] Pardalos, P.M. *Integer and separable programming techniques for large scale global optimization problems.* PhD thesis, Computer Sc. Dept., University of Minnesota (1985).

[ROSE86] Rosen, J.B. and Pardalos, P.M. *Global minimization of large-scale constrained concave quadratic problems by separable programming.* Math. Progr. 34 (1986), 163-174.

[SOUK84] Soukup, J. *Circuit Layout. Proceedings of the IEEE.* Vol. 69, No. 10 (1984), 1281-1304.

[THAK78] Thakur, L.S. *Error Analysis for Convex Separable Programs: The Piecewise Linear Approximation and the Bounds on the Optimal Objective Value.* SIAM J. Appl. Math Vol. 34, No. 4 (1978), 704-714.

[THOA83] Thoai, N.V. and Tuy, H. *Solving the Linear Complementarity Problem through Concave Programming*. Zh. Vychisl. Mat. i. Mat. Fiz. 23(3) (1983), 602-608.

[TUY84] Tuy, H. *Global Minimization of the Difference of two Convex functions.* In: Selected Topics in Operations Research and Mathematical Economics. Lecture Notes Econ. Math. Syst. 226 (1984), 98-118.

[WATA84] Watanabe, H. IC *Layout Generation and Compaction Using Mathematical Optimization*. Ph.D thesis Comp. Sc. Dept. Rochester Univ. (1984).

Chapter 10 Test problems for global nonconvex quadratic programming algorithms

10.1 Introduction

The large-scale global optimization problem with a nonconvex quadratic objective function has the following form:

$$\text{global} \min_{x \in P} f(x,y) = c^T x + \frac{1}{2} x^T Q x + d^T y \qquad \text{(GOP)}$$

$$\text{s.t. } x \in P = \{(x,y) \colon A_1 x + A_2 y \le b, \ x,y \ge 0\} \subseteq R^{n+k}$$

where Q is a symmetric $n \times n$ matrix, $A_1 \in R^{m \times n}$, $A_2 \in R^{m \times k}$, $c, x \in R^n$ and $d, y \in R^k$ (in general $k \gg n$).

Some of the proposed algorithms for (GOP) have been implemented and tested on certain problems. The process of evaluating such implementations is difficult since it requires a variety of test problems with a known global optimum.

The generation of nontrivial test problems for global nonconvex quadratic programming algorithms seems to be difficult, since very few papers have been devoted to that problem. Methods for automatic test problem generators have been proposed by [SUNG82], [KALA86], [PARD87] and [PARD86]. Automatic generation schemes provide a supplement to reference collections of optimization problems.

If Q has at least one negative eigenvalue then there exists a global optimum on the boundary of the (bounded) polyhedron P. When all eigenvalues of Q are nonpositive, i.e. the objective function is concave, then the global minimum occurs at some vertex of P. This property is essential to all methods proposed for test problem generation.

10.2 Global minimum concave quadratic test problems

In this section we consider the following problem:
Given a bounded polyhedron $P = \{x : Ax \le \overline{b}\}$ in R^n and v a nondegenerate vertex,

find a quadratic concave function $f(x)$ whose global minimum over P occurs at v.

A method that generates test problems with a concave quadratic objective function is described in [SUNG82]

Let $A \in R^{m \times n}$ and $m \geq n+1$. Without loss of generality assume that v is determined by the first n hyperplanes that define P, that is, v is given by the system $Bx=b$, where B consists of the first n rows of A and $b \in R^n$. Next solve the multiple-cost-row linear program

$$\min_{x \in P} \sum_{j=1}^{n} a_{ij} x_j, \quad i=1, \ldots, n$$

and denote by c_i the minimum function values. Let $r = (b+c)/2$.

Theorem 10.2.1: [SUNG82] The concave quadratic function $f(x) = -\| Bx - r \|^2$ attains its global minimum over P at v.

Note: To guarantee uniqueness choose $f(x) = -\sum_{i=1}^{n} \omega_i (Bx-r)_i^2$ for arbitrary weights $\omega_i > 0$ and $r_i = (b_i+c_i-\varepsilon)/2$ where $\varepsilon = \min_{1 \leq i \leq n} \{ \frac{\delta}{(b_i-c_i)} \}$, $\delta > 0$.

Another method that generates concave quadratic test problems with a known global minimum can be found in [KALA86].

In [KALA86], the generation of test problems of the following form is considered:

$$\text{global} \min_{x \in P} f(x,y) = -\frac{1}{2}(x-\bar{x})^T Q (x-\bar{x}) + d^T y$$

where P is a bounded polyhedron in R^{n+k}, and $Q_{n \times n}$ is a symmetric positive definite matrix.

Initially test problems are constructed for the case where the objective function is $f(x) = -\frac{1}{2}(x-\bar{x})^T Q (x-\bar{x})$ and $P = \{x : Ax \leq b\}$ is a bounded nonempty polyhedron in R^n. If v_0 is a nondegenerate vertex of P, construct the simplex K through v_0 that contains P. If $K = [v_0, x_1, \ldots, x_n]$, the n vectors $x_1, \ldots, x_n$ lie on the rays emanating from x_0.

Theorem 10.2.2: Given P, $v_0 \in P$ and Q, there exists an $\bar{x} \in R^n$ such that the function $f(x) = -\frac{1}{2}(x-\bar{x})^T Q (x-\bar{x})$ has the property

$$f(x_i) = f(v_0), \quad i=1, \ldots, n.$$

Therefore $\min_{x \in K} f(x) = \min_{x \in P} f(x) = f(v_0)$. Extensions of the same ideas are discussed in [KALA84] for the construction of more general test problems.

10.3 Generation of test problems with nonconvex (concave or indefinite) quadratic objective function

In [PARD86, 87], a new method has been proposed for generation of general nonconvex (not necessarily concave) quadratic test problems. The problem here has the form

$$\text{global} \min_{x \in P} f(x) = c^T x + \frac{1}{2} x^T Q x$$

where $P = \{x : Ax \leq b\}$ is a bounded polyhedron. Given any nondegenerate vertex $v_0 \in P$ and any nonpositive semidefinite matrix, we can find vectors $c \in R^n$ such that the global minimum of $f(x)$ is attained at v_0.

The proposed method initially constructs a simplex K through v_0 that contains P. Then a sphere S is constructed in which K is inscribed. Assume that the center of S is at the origin (i.e. $S = S_0(R)$). Given P, $v_0 \in P$, and any symmetric nonpositive semidefinite matrix, the proposed method finds c such that the global minimum of $f(x) = c^T + 1/2 x^T Q x$ is attained at v_0. The choice of the vector c is motivated by the next theorem.

Theorem 10.3.1: Let $f(x) = c^T x + \frac{1}{2} x^T Q x$, $v_0 \in R^n$ such that $\| v_0 \| = R$, and $c = -(Q + \mu I) v_0$ where $Q + \mu I$ is a positive semidefinite matrix for some $\mu > 0$. Then

$$\min_{\| x \| \leq R} f(x) = f(v_0)$$

Since $P \subseteq K \subseteq S_0(R)$ we have that $\min_{x \in P} f(x) = f(v_0)$.

In [PARD86] another method is presented, using linear complementarity techniques, for the generation of indefinite quadratic problems. With this approach a polyhedral set is defined with a known vertex that is going to be the global minimum of an associated quadratic function (see Exercise 1.).

Small size test problems can be found in several references. The proposed methods generate test problems whose global optimum is attained at some vertex of the polyhedron. Although this is the case with concave problems, for the general indefinite quadratic problem the global optimum may occur at any boundary point. An effort is needed for the generation of indefinite quadratic test problems with nonextreme point solutions.

Regarding different issues on software testing and the reporting of computational results on test problems see [CROW78], [CROW79].

10.4. Exercises

1. Consider the polytope $S(q) = \{x : Mx+q \geq 0,\ x \geq 0\}$, where M is a fixed $n \times n$ matrix and $q \in R^n$.
   Determine a vector q and a vertex $v_0 \in S(q)$ such that
   $$\text{global} \min_{x \in S(q)} f(x) = q^T x + \frac{1}{2} x^T (M+M^T) x$$
   has an optimal solution at v_0 with $f(v_0) = 0$.

2. Given a concave function $f(x)$ (not necessarily quadratic) over R^n, construct a polytope P such that
   $$\text{global} \min_{x \in P} f(x) = f(z)$$
   where z is a given vertex of P.
   Hint: If $\bar{z}$ is the unconstrained maximum of $f(x)$ then chose $z \neq \bar{z}$ and compute $\nabla f(z)$. Randomly choose n directions $d_i \neq 0$ such that $\nabla f(z)^T d_i > 0$ and go along d_i to w_i where $f(w_i) = f(z)$. The points $z, w_1, \ldots, w_n$ form a simplex

whose vertices $z_1, w_1, \ldots, w_n$ are global minima of $f(x)$. Construct nontrivial problems using cutting hyperplanes which eliminate vertices other that z.

10.5 References

[CROW78] Crowder, H., Dembo, R.S., and Mulvey, J.M. *Reporting computational experiments in mathematical programming.* Math. Progr. 15 (1978), 316-329.

[CROW79] Crowder, H., Dembo, R.S., and Mulvey, J.M. *On reporting computational experiments with mathematical software.* ACM Trans. on Mathem. Soft. 5 (1979), 193-203.

[KALA86] Kalantari, B. and Rosen, J.B. *Construction of large-scale global minimum concave quadratic test problems.* JOTA 48 (1986), 303-313.

[PARD86] Pardalos, P.M. *On generating test problems for global optimization algorithms problem.* Tech. Report CS-86-01, Comput. Sci. Dept., Penn. State University.

[PARD87] Pardalos, P.M. *Generation of large-scale quadratic programs for use as global optimization test problems.* To appear in ACM Trans. on Math. Software (1987).

[SUNG82] Sung, Y.Y. and Rosen, J.B. *Global minimum test problem construction.* Math. Progr. 24 (1982), 353-355.

General References of Constrained Global Optimization

The following bibliographic list contains references dealing mainly with deterministic approaches for nonconvex problems. Although most of the papers consider the concave or indefinite quadratic problem, some references are concerned with a variety of applications and related problems.

[AL-K83] Al-Khayyal, F.A. and Falk, J.E. *Jointly constrained biconvex programming.* Math. of Oper. Research Vol. 8, No. 2 (1983), 273-286.

[AL-K83] Al-Khayyal, F.A. *Jointly constrained bilinear programming and related problems.* Industrial Systems Engineering Report Series No. C-83-3, The Georgia Inst. of Technology, Atlanta, GA.

[ALTM68] Altman, M. *Bilinear programming.* Bull. Acad. Polon. Sci. Ser. Sci. Math. Astronom. Phys.16 (1968), 741-745.

[ALUF85] Aluffi-Pentini, F., Parisi, V., and Zirilli, F. *Global optimization and stochastic differential equations.* JOTA Vol. 47, No. 1 (1985), 1-17.

[ANEJ84] Aneja, Y.P, Aggarwal, V., and Nair, K.P.K. *On a class of quadratic programs.* Europ. Journ. of Oper. Res. 18 (1984), 62-72.

[ARRO61] Arrow, K.J. and Enthoven, A.D. *Quasiconcave programming.* Econometrica 29 (1961), 779-800.

[ARSH81] Arshad, M. and Khan, A. *A procedure for solving concave quadratic problems.* Aligarh J. Statist.1, No.2 (1981), 106-112.

[AVRI76] Avriel, M. *Nonlinear Programming: Analysis and Methods.* Prentice-Hall Inc., Englewood Cliffs, N.J. (1976).

[BACH78] Bachem, A. and Korte, B. *An algorithm for quadratic optimization over transportation polytopes.* Z. Angew. Math. Mech. 58 (1978), 459-461.

[BALA75] Balas, E. *Nonconvex quadratic programming via generalized polars.* SIAM J. Appl. Math. 28 (1975), 335-349.

[BALA73] Balas, E. and Burdet, C.A. *Maximizing a convex quadratic function subject to linear constraints.* MSRR 299, GSIA (1973), Carnegie-Mellon Univ.

[BALI73] Bali, S. *Minimization of a concave function on a bounded convex polyhedron.* Ph.D.Dissertation Univ. of California (1973), Los Angeles.

[BALI61] Balinski, M.L. *An algorithm for finding all vertices of polyhedral sets.* SIAM Jour. Vol.9, No.1 (1961), 72-89.

[BANS75A] Bansal, P.P and Jacobsen, S.E. *Characterization of local solutions for a class of nonconvex problems.* JOTA Vol.15, No.5 (1975), 549-564.

[BANS75B] Bansal, P.P and Jacobsen, S.E. *An algorithm for optimizing network flow capacity under economies of scale.* JOTA Vol.15, No.5 (1975), 565-586.

[BARD82] Bard, J.F. and Falk, J.E. *A separable programming approach to the linear complementarity problem.* Comput. and Ops Res. Vol.9, No.2 (1982), 153-159.

[BARR81] Barr, R.S., Glover, F. and Klingman, D. *A new optimization method for large scale fixed charge transportation problems.* Oper. Res. 29, No. 3 (1981), 448-463.

[BASS85] Basso, P. *Optimal search for the global maximum of functions with bounded seminorm.* SIAM J. Numer. Anal. Vo. 22, No. 5 (1985), 888-903.

[BASS82] Basso, P. *Iterative methods for the localization of the global maximum.* SIAM J. Numer. Anal. 19, No. 4 (1982), 781-792.

[BASS85] Basso, P. *On effectiveness measures for optimal search measures.* To appear in Lecture Notes in Control and Optimization.

[BAZA82] Bazara, M.S. and Sherali, H.D. *On the use of exact and heuristic cutting plane methods for the quadratic assignment problem.* J. Oper. Res. Soc. Vol.13 (1982), 991-1003.

[BEAL80] Beale, E.M.L. *Branch and bound methods for numerical optimization of nonconvex functions*. Compstat 1980 (Proc.Fourth Symp. Comput. Stat., Edinburgh, 1980), pp.11-20, Physica, Vienna.

[BEAL76] Beale, E.M.L. and Forrest, J.J.H. *Global optimization using special ordered sets*. Math. Progr.10 (1976), 52-69.

[BEAL78] Beale, E.M.L. and Forrest, J.J.H. *Global optimization as an extension of integer programming*. Towards Global Optimisation 2, L.C.W. Dixon and G.P.Szego (eds) North Holland Publ. Comp. (1978), 131-149.

[BEAL70] Beale, E.M.L. and Tomlin, J.A. *Special facilities in a general mathematical programming system for nonconvex problems using ordered sets of variables*. In Proceedings of the Fifth international conference on Oper. Res., Ed. J. Lawrence, 1970 (Tavistock Publications, London) 447-454.

[BENA85] Benacer, R. and Pham, D.T. *Global maximization of a nondefinite quadratic function over a convex polyhedron*. FERMAT Days 1985: Mathematics for optimization, J.-Hiriart-Urruty (editor) Elsevier Sci. Publishers, 65-76.

[BEND62] Benders, J.F. *Partitioning procedures for solving mixed-variables programming problems*. Numer. Math. 4 (1962), 238-252.

[BENS85] Benson, H.P. *A finite algorithm for concave minimization over a polyhedron*. Naval Res. Log. Quarterly, Vol. 32 (1985), 165-177.

[BENS82A] Benson, H.P. *On the convergence of two branch and bound algorithms for non convex programming problems*. JOTA Vol.36, No.1 (1982), 129-134.

[BENS82B] Benson, H.P. *Algorithms for parametric nonconvex programming*. JOTA Vol.38, No.3 (1982), 316-340.

[BERT79] Bertsekas, D.P. *Convexification procedures and decomposition methods for nonconvex optimization problems*. JOTA Vol.29 (1979), 169-197.

[BHAT81] Bhatia, H.L. *Indefinite quadratic solid transportation problem*. *J. of Information and Optimization*. Vol. 2, No. 3 (1981), 297-303.

[BOEN82] Boender, C.G.E. and Rinnooy Kan, A.H.G., Stougie, L., and Timmer, G.T. *A stochastic method for global optimization.* Math. Progr. 22 (1982), 125-140.

[BULA82] Bullatov, V.P. and Kasinskaya, L.I. *Some methods of concave minimization on a convex polyhedron and their applications* (Russian). Methods of Optimization and their applications, "Nauka" Sibirsk. Otdel, Novosibirsk (1982), 71-80.

[BURD74] Burdet, C.A. *Generating all the faces of a polyhedron.* SIAM Jour. Vol.26, No.1 (1974), 72-89.

[BURD70] Burdet, C.A. *Deux modeles de minimisation d'une fonction economique concave.* R.I.R.O. Vol. 1 (1970), 79-84.

[BYRD86] Byrd, R.H., Dert, C.L., Rinnooy Kan, A.H.G. and Schnabel, R.B. *Concurrent stochastic methods for global optimization.* Technical Report CU-CS-338-86, Dept. of Computer Sci., Univ. of Colorado at Boulder.

[CABO74] Cabot, A.V. *Variations on a cutting plane method for solving concave minimization problems with linear constraints.* Naval Res. Logist. Quart. 21 (1974), 265-274.

[CABO70] Cabot, A.V. and Francis, R.L. *Solving nonconvex quadratic minimization problems by ranking the extreme points.* Oper. Res.18 (1970), 82-86.

[CAND64] Candler, W. and Townsley, R.J. *The maximization of a quadratic function of variables subject to linear inequalities.* Manag. Sc. Vol 10, No. 3, 515-523.

[CARI77] Carillo, J.J. *A relaxation algorithm for the minimization of a quasiconcave function on a convex polyhedron.* Math. Progr. 13 (1977), 69-80.

[CARV72] Carvajal-Morero, R. *Minimization of concave functions subject to linear constraints.* Oper. Res. Center, Univ. of California, Berkeley, ORC 72-3 (1972).

[CHUN81] Chung, S.J. and Murty, K.G. *Polynomially bounded ellipsoid algorithms for convex quadratic programming.* In O.L. Mangasarian,

R.R. Meyer and S.M. Robinson (editors): Nonlinear Programming 4, Academic Press, 1981, 439-485.

[CIRI85] Cirina, M. *A class of nonlinear programming test problems.* Working paper (1985), Dip. di Informat. Torino, Italy.

[CIRI86] Cirina, M. *A finite algorithm for global quadratic minimization.* Working paper (1986), Dip. di Informat. Torino, Italy.

[CALV86] Calvert, B. and Vamanamurthy, M.K. *Local and global extrema for functions of several variables.* J. Austral. Math. Soc. (A) 29 (1986), 362-368.

[CAND64] Candler, W. and Townsley, R.J. *The maximization of a quadratic function of variables subject to linear inequalities.* Manag. Science, Vol. 10, No. 3 (1964), 515-523.

[COTT70] Cottle, R. and Mylander, W.C. *Ritter's cutting plane method for nonconvex quadratic programming.* In: J. Abadie, ed. Integer and Nonlinear Programming. North-Holland (1970), Amsterdam.

[CZOC82A] Czochralska, I. *Bilinear programming.* Zastosow. Mat.17 (1982), 495-514.

[CZOC82B] Czochralska, I. *The method of bilinear programming for nonconvex quadratic problems.* Zastosow . Mat.17 (1982), 515-525.

[DIEW81] Diewert, W.E., Avriel M., and Zang, I. *Nine kinds of quasiconcavity and concavity.* Journal of Econ. Theory 25 (1981), 397-420.

[DIXO75] Dixon, L.C.W. and Szego, G.P. eds., *Towards global optimisation.* (North-Holland, Amsterdam, 1975).

[DIXO78] Dixon, L.C.W. and Szego, G.P. eds., *Towards global optimisation 2.* (North-Holland, Amsterdam, 1978).

[DYER77] Dyer, M.E. and Proll, L.G. *An algorithm for determining all extreme points of a convex polytope.* Math. Progr. 12 (1977), 81-96.

[DYER83] Dyer, M.E. *The complexity of vertex enumeration methods.* Math. Oper. Res. 8 (1983), 381-402.

[ERIC85] Erickson, R.E., Monna, C.L. and Veinott, A.F. *Minimum concave cost network flows.* To appear in Math. of Oper. Res.

[EVER63] Everett, H.III. *Generalized Lagrange multiplier method for solving problems of optimum allocation of resources.* Oper. Res.11 (1963), 399-417.

[EVTU85] Evtushenko, G.Y. *Numerical Optimization Techniques.*(Translation series in Math. and Engin.) Optimization Software Inc., Publications Division N.Y. (1985)

[FALK72] Falk, J.E. *An algorithm for locating approximate global solutions of nonconvex, separable problems.* Technical paper Serial T-262 (1972). Program in Logistics. The George Washington Un.

[FALK73] Falk, J.E. *A linear max-min problem.* Math. Progr. 5 (1973), 169-188.

[FALK69] Falk, J.E. *Lagrange multipliers and nonconvex programs.* SIAM Journal of Control 7 (1969), 534-545.

[FALK76] Falk, J.E. and Hoffman, K.L. *A successive underestimating method for concave minimization problems.* Math. of Oper. Res.1 (1976), 251-259.

[FALK80] Falk, J.E. and Hoffman, K.L. *Concave minimization via collapsing polytopes.* Technical Paper Serial T-438 (1980), Inst. for Manag. Sci. and Eng., George Washington University.

[FALK69] Falk, J.E. and Soland, R.M. *An algorithm for separable nonconvex programming problems.* Manag. Sc. Vol.15, No.9 (1969), 550-569.

[FLOR86] Florian, M. *Nonlinear cost network models in transportation analysis.* Math. Progrm. Study 26 (1986), 167-196.

[FLOR71] Florian, M. and Robillard, P. *An implicit enumeration algorithm for the concave cost network flow problem.* Manag. Sci. Vol.18, No.3 (1971), 184-193.

[FLOR71] Florian, M., Rossin-Arthiat, M., and de Verra, D. *A property of minimum concave cost flows in capacited networks.* Canadian Journal of Oper. Res. 9 (1971), 293-304.

[FORG72] Forgo, F. *Cutting plane methods for solving nonconvex quadratic problems.* Acta Cybernet. Vol. 1 (1972), 171-192.

[FRIE74] Frieze, A.M. *A bilinear programming formulation of the 3-dimensional assignment problem.* Math. Progr.7 (1974), 376-379.

[FUJI85] Fujiwara, O. *Notes on differentiability of global optimal values.* To appear in Math. of Oper. Res.

[FULO83] Fulop, J. *Finding the original vertex and its application minimizing a concave function subject to linear constraints.* Alkalmaz. Mat. Lapok.9, No. 1-2 (1983), 51-72.

[GALL80] Gallo, G., Sandi, C. and Sodini, C. *An algorithm for the min concave cost flow problem.* European J. of Oper. Res. 4 (1980), 249-255.

[GALL79] Gallo, G. and Sodini, C. *Adjacent extreme flows and application to min concave cost flow problems.* Networks 9 (1979), 95-122.

[GALL77] Gallo, G. and Ulculu, A. *Bilinear Programming: an exact algorithm.* Math. Progr.12 (1977), 173-174.

[GALP85] Galperin, E.A. *The cubic algorithm.* Journ. of Math. Anal. and Appl. 112 (1985), 635-640.

[GALP85] Galperin, E.A. and Zheng, Q. *Nonlinear observation via global optimization: The measure theory approach.* To appear in JOTA.

[GARE76] Garey, M.R., Johnson, D.S. and Stockmeyer, L. *Some simplified NP-complete problems.* Theor. Comp. Sc.1 (1976), 237-268.

[GASA84] Gasanov, I.I. and Rikun, A.D. *On necessary and sufficient conditions for uniextremality in nonconvex mathematical programming problems.* Soviet Math. Dokl. Vol. 30, No. 2 (1984), 457-459.

[GE87] Ge, R.P. *The theory of filled function method for finding global minimizers of nonlinearly constrained minimization problems.* J. of Computational Math., Vol. 5, No. 1 (1987), 1-9.

[GIAN76] Giannessi, F. and Niccolucci, F. *Connections between nonlinear and integer programming problems.* Symposia Mathematica Vol. XIX, Istituto Nazionale Di Alta Math., Acad. Press N.Y. (1976), 161-176

[GINS73] Ginsberg, W. *Concavity and quasiconcavity in economics.* Journ. of Econ. Theory 6 (1973), 596-605.

[GLOV73A] Glover, F. *Convexity cuts and cut search.* Oper.Res.21, No.1 (1973), 123-134.

[GLOV73B] Glover, F. and Klingman, D. *Concave programming applied to a special class of 0-1 integer programs.* Oper. Res.21, No. 1 (1973), 135-140.

[GFRE84] Gfrerer, H. *Globally convergent decomposition methods for nonconvex optimization problems.* Computing 32 (1984), 199-227.

[GOME81] Gomez, S. and Levy, A.V. *The tunneling method for solving the constrained global optimization problem with several nonconnected feasible regions.* Springer-Verlag, Lecture Notes in Math., No. 909 (1981).

[GRAV85] Graves, S.T. and Orlin, J.B. *A minimum concave-cost dynamic network flow problem with an application to lot-sizing.* Networks, Vol. 15 (1985), 59-71.

[GROT76] Grotte, J.H. *Program MOGG-A code for solving separable non convex optimization problems.* P-1318 (1976). The Institute for Defense Analysis, Arlington, Virginia.

[GROT85] Grotzinger, S.I. *Supports and convex envelopes.* Math. Progr. 31 (1985), 339-347.

[GUPT83] Gupta, A.K. and Sharma, J.K. *A generalized simplex technique for solving quadratic programming problems.* Indian Journ. of Techn. 21 (1983), 198-201.

[HADL64] Hadley, G. *Nonlinear and Dynamic Programming.* Addison-Wessley Publ. Co. (1964).

[HAMM65] Hammer (Ivanescu), P.L. *Some network flow problems solved with pseudo- Boolean programming.* Oper. Res. Vol.13 (1965), 388-399.

[HANS80] Hansen, E. *Global optimization using interval analysis - the multidimensional case.* Numerische Math. 34 (1980), 247-270.

[HEIS81] Heising-Goodman, C.D. *A survey of methodology for the global minimization of concave functions subject to convex constraints.* OMEGA, Intl.Jl. of Mgmt. Sci. 9 (1981), 313-319.

[HIRS61] Hirsch, W.M. and Hoffman, A.J. *Extreme varieties, concave functions, and the fixed charge problem.* Comm. Pure Appl. Math. XIV (1961), 355-369.

[HOFF75] Hoffman, K.L. *A successive underestimating method for concave minimization problems.* Ph.D.thesis (1975), The George Washington University

[HOFF81] Hoffman, K.L. *A method for globally minimizing concave functions over convex sets.* Math. Progr.20 (1981), 22-32.

[HORS76A] Horst, R. *An algorithm for nonconvex programming problems.* Math. Progr.10 (1976), 312-321.

[HORS76B] Horst, R. *A new branch and bound approach for concave minimization problems.* Lecture Notes in Computer Sc., Vol. 41 (1976), 330-337.

[HORS78] Horst, R. *A new approach for separable nonconvex minimization problems including a method for finding the global minimum of a function of a single variable.* Proceeding in Oper. Res., 7 (Sixth Annual Meeting, Deutsche Gesellsch. Oper. Res., Christian-Albrechts-Univ.Kiel 1977) 39-47, Physica, Würzburg 1978.

[HORS80] Horst, R. *A note on the convergence of an algorithm for nonconvex programming problems.* Math. Progr. 19 (1980), 37-238.

[HORS82] Horst, R. *A note on functions, whose local minimum are global.* JOTA Vol.36, No.3 (1982), 457-463.

[HORS84A] Horst, R. *On the convexification of nonlinear programming problems: An applications oriented survey.* European J.of Oper. Res.15 (1984), 382-392.

[HORS84B] Horst, R. *On global minimization of concave functions: Introduction and Survey.* Operations Research Spektrum, Vol. 6 (1984), 195-205.

[HORS85] Horst, R. *A general class of branch and bound methods in global optimization.* To appear in JOTA.

[HU69] Hu, T.C. *Minimizing a concave function in a convex polytope*. MRC Technical Summary Report No.1011 (1969), Math. Research Center, Un.of Wisconsin.

[ISTO77] Istomin, L.A. *A modification of Tuy's method for minimizing a concave function over a polytope*. Zh. Vychisl. Mat. mat. Fiz., Vol. 17 (1977), 1582-1592.

[IVAN76] Ivanilov, Yu I. and Mukhamediev, B.M. *An algorithm for solving the linear max-min problem*. Izv. Akad. Nauk SSSR, Tekhn. Kibernitika, No.6 (1976), 3-10.

[JACO81] Jacobsen, S.E. *Convergence of a Tuy-type algorithm for concave minimization subject to linear inequality constraints*. Appl. Math. Opt.7 (1981), 1-9.

[JONG77] Jongen, H.T. *On non-convex optimization*. Dissertation, Twente University of Technology (1977), The Netherlands.

[KALA82] Kalantari, B.and Rosen, J.B. *Penalty for zero-one integer equivalent problems*. Math. Progr.24 (1982), 229-232.

[KALA84A] Kalantari, B. *Large scale global minimization of linearly constrained concave quadratic functions and related problems*. Ph.D.thesis, Computer Sc.Dept. (1984), University of Minnesota.

[KALA86] Kalantari, B.and J.B. Rosen, J.B. *Construction of large-scale global minimum concave quadratic test problems*. JOTA 48 (1986), 303-313.

[KALA85] Kalantari, B, and Rosen, J.B. *An algorithm for large-scale global minimization of linearly constrained concave quadratic functions*. TR-147 (1985), Dept. of Computer Sc., Rutgers Univ.

[KEDE83] Kedem, G. and Watanabe, H. *Optimization techniques for IC layout and compaction*. Proceedings IEEE Intern. Conf. in Computer Design: VLSI in Computers (1983), 709-713.

[KELL60] Kelley, J.E.Jr. *The cutting plane method for solving convex programs*. J. Soc. Indust. Appl. Math.8 (1960), 703-712.

[KLEI67] Kleibohm, K. *Bemerkungen zum problem der nichtkonvex programmierung.* (Remarks on the nonconvex programming problem). Unternehmensf. 11, 49-60.

[KONN76A] Konno, H. *A cutting plane algorithm for solving bilinear programs.* Math. Progr.11 (1976), 14-27.

[KONN76B] Konno, H. *Maximization of a convex quadratic function subject to linear constraints.* Math. Progr.11 (1976), 117-127.

[KONN80] Konno, H. *Maximizing a convex quadratic function over a hypercube.* J. Oper. Res. Soc. Japan 23 No.2 (1980), 171-189.

[KONN81] Konno, H. *An algorithm for solving bilinear knapsack problems.* J. Oper. Res. Soc. Japan 24, No.4 (1981), 360-373.

[KORT67] Kortanek, K.O. and Eaves, J.P. *Pseudoconcave programming and Lagrange regularity.* Oper. Res. Vol. 15 (1967), No. 6.

[KOUG79] Kough, P.L. *The indefinite quadratic programming problem.* Oper. Res. Vol. 27, No. 3 (1979), 516-533.

[KOZL79] Kozlov, M.K.,Tarasov, S.P., and Khachian, L.G. *Polynomial solvability of convex quadratic programming.* Soviet Math. Dokl. Vol.20, No.5 (1979), 1108-1111.

[KRYN79] Krynski, S.L. *Minimization of a concave function under linear constraints (modification of Tuy's method).* Survey of Math. Progr. (Proc. Ninth Intrn. Math. Progr. Sympos., Budapest 1976) Vol.1 (1979), 479-493. North-Holland, Amsterdam 1979

[LAWL63] Lawler, E.L. *The quadratic assignment problem.* Manag. Sc.9 (1963), 586-599.

[LEBE82] Lebedev, V.Yu. *A method of solution of the problem of maximization of a positive definite quadratic form on a polyhedron.* Zh. Vychisl. Mat. i. Mat. Fiz.22, No.6 (1982), 1344-1351.

[LEVY85] Levy, A.V. and Gomez, S. *The tunneling algorithm for global minimization of functions.* To appear in SIAM J. Sci. Stat. Comput. (1985).

[LIN86] Lin, L.S. and Allen, J. *Minplex- A compactor that minimizes the bounding rectangle and individual rectangles in a layout.* 23rd Design Automation Conference (1986), 123-129.

[MAJT74] Majthay, A. and Whinston, A. *Quasiconcave minimization subject to linear constraints.* Discrete Math. 1 (1974), 35-39.

[MALI82] Maling, K., Mueller, S.H., and Heller, W.R. *On finding most optimal rectangular package plans.* Proceeding of the 19th Design Automation Conference (1982), 663-670.

[MANA74] Manas, M. and Nedoma, J. *Finding all vertices of a convex polyhedral set.* Numerische Mathem. Vol.9, No.1 (1974), 35-59.

[MANA68] Manas, M. and Nedoma, J. *Finding all vertices of a convex polyhedron.* Numerische Mathem. 12 (1968), 226-229.

[MANA68] Manas, M. *An algorithm for a nonconvex programming problem.* Econom. Mathem. Obzor 4(2) (1968), 202-212.

[MANC76] Mancini, L. and McCormick, G.P. *Bounding global minima.* Math. of Operations Res., Vol.1, No. 1 (1976), 50-53.

[MANG84] Mangasarian, O.L. and Shiau, T.H. *A variable complexity norm maximization problem.* Computer Sc. Dept. University of Wisconsin Tech. Report #568 (1984).

[MANG78] Mangasarian, O.L. *Characterization of linear complementarity problems as linear programs.* Math. Progr. Study 7 (1978), 74-87.

[MART71] Martos, B. *Quadratic Programming with a quasiconvex objective function.* Oper. Res. 19 (1971), 87-97.

[MATH80] Matheiss, T.H. and Rubin, D.S. *A survey and comparison of methods for finding all vertices of convex polyhedral sets.* Math. of Oper .Res. Vol.5, No.2 (1980), 167-185.

[MATH73] Mattheis, T.H. *An algorithm for the determination of unrelevant constraints in systems of linear inequalities.* Oper. Res. Vol. XXI (1973), 247-260

[McCO72A] McCormick, G.P. *Converting general nonlinear programming problems to separable nonlinear programming problems.* Technical

Paper Serial T-267 (1972). Program in Logistics, George Wash. Univ. Wash. DC.

[McCO72B] McCormick, G.P. *Attempts to calculate global solutions of problems that may have local minima.* Numerical Methods for Non-linear Optimization F.A.Lootsma ed. Acad. Press 1972, 209-221.

[McCO76] McCormick, G.P. *Computability of global solutions to factorable nonconvex programs: Part I-convex underestimating problems.* Math. Progr.10 (1976), 147-175.

[McCO83] McCormick, G.P. *Nonlinear Programming: Theory, Algorithms and Applications.* John Wiley and Sons, NY 1983.

[McKE78] McKeown, P.G. *Extreme point ranking algorithms: A computational survey.* Computers and Math. Progr., W.W.White ed. National Bureau of Standards Special Publication 502, U.S. Government Printing Office, Washington DC 20402, 1978, 216-222.

[MEYE83] Meyer, R.R. *Computational aspects of two-segment separable programming.* Math. Progr. 26 (1983), 21-39.

[MEYE76] Meyer, R.R. *Mixed integer minimization models for piecewise linear functions of a single variable.* Discrete Math. 16 (1976), 163-171.

[MLAD86] Mladineo, R.H. *An algorithm for finding the global maximum of a multimodal, multivariate function.* Math. Program. 34 (1986), 188-200.

[MUEL70] Mueller, R.K. *A method for solving the indefinite quadratic programming problem.* Manag. Sci. 16 (1970), 333-339.

[MUKH82] Mukhamediev, B.M. *Approximate methods of solving concave programming problems.* Zh. Vychisl. Mat.i. Mat. Fiz.22, 727-732 (translation: USSR Comput. Maths. Math. Phys. Vol.22, No.3 (1982), 238-245).

[MURT69] Murty, K. *Solving the fixed charge problem by ranking the extreme points.* Oper. Res.16 (1969), 268-279.

[PARD85] Pardalos, P.M. *Integer and Separable programming techniques for large-scale global optimization problems.* Ph.D. thesis (1985),

Computer Science Department, Univ. Minnesota, Minneapolis MN.

[PARD85] Pardalos, P.M., Glick, J.H., and Rosen, J.B. *Global minimization of indefinite quadratic problems.* Technical Report (1985), Computer Sci. Dept. The Pennsylvania State University.

[PARD85] Pardalos, P.M. *On generating test problems for global optimization algorithms.* Technical Report, Computer Sci. Dept. The Pennsylvania State Univesity.

[PARD86A] Pardalos, P.M. *Enumerative techniques for solving some nonconvex global optimization problems.* Technical Report CS-86-33, Computer Sci. Dept., The Pennsylvania State Univ.

[PARD86B] Pardalos, P.M. *Aspects of parallel computation in global optimization.* Proc. of the 24th Annual Allerton Conference on Communication, Control and Computing (1986), 812-821.

[PARD86C] Pardalos, P.M. and Rosen, J.B. *Methods for global concave minimization: A bibliographic survey.* SIAM Review, Vol. 28, No. 3 (1986), 367-379.

[PARD87A] Pardalos, P.M. and Kovoor, N. *An algorithm for singly constrained quadratic programs.* Technical Report, Computer Sci. Dept. (1987), Penn. State Univ.

[PARD87B] Pardalos, P.M. and Schnitger, G. *Checking local optimality in constrained quadratic programming is NP-hard.* Technical Report, Computer Sci. Dept. (1987), Penn. State Univ.

[PARD87C] Pardalos, P.M. *Generation of large-scale quadratic programs for use as global optimization test problems.* To appear in ACM Trans. om Math. Software.

[PARD87D] Pardalos, P.M. *Quadratic programming defined on a convex hull of points.* Technical Report, Computer Sci. Dept. (1987), Penn. State Univ.

[PARD87E] Pardalos, P.M. *Objective function approximation in nonconvex programming.* Proceedings of the 18th Modeling and Simulation Conf. (1987).

[PARD87F] Pardalos, P.M. and Rosen, J.B. *Global optimization approach to the linear complementarity problem.* To appear in SIAM J. Scient. Stat. Computing (1987).

[PASS78] Passy, U. *Global solutions of mathematical programs with intrinsically concave functions.* JOTA Vol.26, No.1 (1978), 97-115.

[PATE83] Patel, N.R. and Smith, R.L. *The asymptotic extreme value distribution of the sample minimum of a concave function under linear constraints.* Oper. Res. Vol. 31, No. 4 (1983), 789-794.

[PHAM84] Pham, Dinh T. *Algorithmes de calcul du maximum de formes quadratiques sur la boule unite de la forme du maximum.* Numer. Math. 45 (1984), 377-401.

[PHAM85] Pham, Dinh T. and Souad, El B. *Algorithms for solving a class of nonconvex optimization problems: Methods of subgradients.* FERMAT Days 1985: Mathematics for optimization, J.-Hiriart-Urruty (editor) Elsevier Sci. Publishers, 249-271.

[PHIL87] Phillips, A. and Rosen, J.B. *Multitasking mathematical programming algorithms.* To appear in Annals of Oper. Res. (1987), special volume on "Parallel Optimization on Novel Computer Architectures".

[PFOR84] Pforr, E.A. and Gunther, R.H. *Die Normalform des quadratischen Optimierungsproblems und die pol-polaren theorie.* Math. Oper. u Stat. Ser. Optim. 15, 1 (1984), 41-55.

[PIJA72] Pijavski, S.A. *An algorithm for finding the absolute extremum of a function.* USSR Comput. Math. and Math. Phys. (1972), 57-67.

[PINT86] Pinter, J. *Extended univariate algorithms for n-dimensional global optimization.* Computing 36 (1986), 91-103.

[PINT86] Pinter, J. *Globally convergent methods for n-dimensional multiextremal optimization.* To appear in: Math. Operationsforsch. u. Stat. Ser. Opt.

[PINT86] Pinter, J. *Global optimization on convex sets.* To appear in: OR Spectrum.

[RAGH69] Raghavachari, M. *On connections between zero-one integer programming and concave programming under linear constraints.* Oper. Res.17 (1969), 680-684.

[RECH70] Rech, P. and Barton, L.G. *A non-convex transportation algorithm.* In: Beale E.M.L. ed., Applications of Mathematical Programming Techniques, American Elsevier N.Y. (1970).

[REEV75] Reeves, G.R. *Global minimization in non-convex all-quadratic programming.* Manag. Sc. Vol.22 (1975), 76-86.

[RINN85] Rinnooy Kan, A.H.G., Boender C.G.E., and Timmer, G.T. *A stochastic approach to global optimization.* Computational Mathematical Programming, (Springer-Verlag (1985), K. Schittkowski, Ed.), 282-308.

[RITT66] Ritter, K. *A method for solving maximum problems with a nonconcave quadratic objective function.* Z. Wahrscheinlichkeitstheorie Geb.4, 340-351.

[RITT65] Ritter, K. *Stationary points of quadratic maximum problems.* Z. Wahrscheinlich. verw. Geb. 4 (1965), 149-158.

[ROCK70] Rockafellar, R.T. *Convex Analysis.* Princeton Univ. Press N.J. (1970).

[ROSE83A] Rosen, J.B. *Global minimization of a linearly constrained concave function by partition of feasible domain.* Math. Oper. Res.8 (1983), 215-230.

[ROSE83B] Rosen, J.B. *Parametric global minimization for large scale problems.* Tech. Rep. 83-11 (revised), Computer Sc. Dept. Univ. of Minnesota.

[ROSE84A] Rosen, J.B. *Performance of approximate algorithms for global minimization.* Math. Progr. Study 22 (1984), 231-236.

[ROSE84B] Rosen, J.B. *Computational solution of large-scale constrained global minimization problems.* Numerical Optimization 1984. (P.T.Boggs, R.H. Byrd, R.B.Schnabel, Eds.) SIAM, Phil., 263-271.

[ROSE86] Rosen, J.B. and Pardalos, P.M. *Global minimization of large-scale constrained concave quadratic problems by separable programming.* Mathem. Program. 34 (1986), 163-174.

[ROSS73] Rossler, M. *A method to calculate an optimal production plan with a concave objective function.* Unternehmensforschung Vol.20, No.1 (1973), 373-382.

[SCHA73] Schaible, S. *On factored quadratic programs.* Zeitschrift für Oper. Res. 17 (1973), 179-181.

[SEN85] Sen, S. and Sherali, H.D. *On the convergence of cutting plane algorithms for a class of nonconvex mathematical programs.* Math. Progr. Vol. 31, No. 1 (1985), 42-56.

[SEN85] Sen, S. and Sherali, H.D. *A branch and bound algorithm for extreme point mathematical programming problems.* Discrete Appl. Math. Vol. 11 (1985), 265-280.

[SHER80A] Sherali, H.D. and Shetty, C.M. *A finitely convergent algorithm for bilinear programing problems using polar and disjunctive face cuts.* Mathm. Progr. Vol. 19 (1980), 14-31.

[SHER80B] Sherali, H.D. and Shetty, C.M. *Deep cuts in disjunctive programming.* Naval Research Log. Quart. Vol. 27 (1980), 453-476.

[SHER80C] Sherali, H.D. and Shetty, C.M. *Optimization with disjunctive constraints.* Springer N.Y. (1980).

[SHER82] Sherali, H.D. and Shetty, C.M. *A finitely convergent procedure for facial disjunctive programs.* Discrete Applied Math. Vol. 4 (1982), 135-148.

[SHER85] Sherali, H. and Sen, S. *A disjunctive cutting plane algorithm for the extreme point mathematical programming problems.* Opsearch (Theory) Vol. 22, No. 2 (1985), 83-94.

[SCHN86] Schnabel, R.B. *Concurrent function evaluations in local and global optimization.* Technical Report CU-CS-345-86, Computer Sci. Dept., Univ. of Colorado at Boulder.

[SHUB72] Shubert, B. *A sequential method seeking the global maximum of a function.* SIAM J. of Num. Analysis, Vol. 9, No. 3 (1972), 379-388.

[SOLA71] Soland, R.M. *An algorithm for separable nonconvex programming problems II: nonconvex constraints.* Manag. Sc.17, No.11 (1971), 759-773.

[SOLA74] Soland, R.M. *Optimal facility location with concave costs.* Oper. Res. (1974), 373-382.

[SUNG82] Sung, Y.Y. and Rosen, J.B. *Global minimum test problem construction.* Math. Progr. 24 (1982), 353-355.

[SWAR66] Swarup, K. *Quadratic programming.* Cahiers du Centre d'Etudes de Res. Oper. 8 (1966), 223-234.

[SWAR66] Swarup, K. *Indefinite quadratic programming.* Cahiers du Centre d'Etudes de Res. Oper. 8 (1966), 217-222.

[TAHA73] Taha, H.A. *Concave minimization over a convex polyhedron.* Naval Res. Logist. Quart.20, No.1 (1973), 533-548.

[THAK85] Thakur, L.S. *Domain contraction in nonconvex programming: Minimizing a quadratic concave objective under linear constraints.* Working paper Series B, #92 (1985), Yale School of Organiz. and Managm.

[THIE80] Thieu, T.V. *Relationship between bilinear programming and concave minimization under linear constraints.* Acta Math. Vietnam 5, No.2 (1980), 106-113.

[THOA81] Thoai, N.V. *Application of the extension principle to solutions of concave optimization problems.* Math. Operationsforsch. and Stat. Ser. Optimiz. Vol.12, No.1 (1981), 45-51.

[THOA80] Thoai, N.V. and Tuy, H. *Convergent algorithms for minimizing a concave function.* Math. of Oper. Res. Vol.5, No.4 (1980), 556-566.

[THOA83] Thoai, N.V. and Tuy, H. *Solving the linear complementarity problem through concave programming.* Zh. Vychisl. Mat. i. Mat. Fiz. 23(3) (1983), 602-608.

[TUY64] Tuy, H. *Concave programming under linear constraints.* Dokl.Akad.Nauk. SSSR, 159, 32-35 (Translated: Soviet Math. Dokl.5 (1964), 1437-1440).

[TUY82] Tuy, H. *Global maximization of a convex function over a closed, convex, not necessarily bounded set.* Cahiers de Mathematiques de la Decision No.8223, Universite Paris, Daupfine 1982.

[TUY83] Tuy, H..*On outer approximation methods for solving concave minimization problems.* Report No.108 (1983), Forschungsschwerpunkt Dynamische Systeme Universität Bremen, West Germany.

[TUY84] Tuy, H. *Global minimization of a difference of two convex functions.* In "Selected Topics in Operations Research and Mathematical Economics", Lecture Notes Econ. Math. Syst. 226 (1984), 98-118.

[TUY85] Tuy, H. *A general deterministic approach to global optimization via d.c. programming.* FERMAT Days 1985: Mathematics for optimization, J. -Hiriart-Urruty (editor) Elsevier Sci. Publishers, 273-303.

[TUY85] Tuy, H. Thieu. T.V. and Thai. N.Q. *A conical algorithm for globally minimizing a concave function over a closed convex set.* Mathem. of Oper. Res. Vol 10, No. 3 (1985), 498-514.

[TUY85] Tuy, H. *Concave minimization under linear constraints with special structure.* Optimization Vol 16. No. 3 (1985), 335-352.

[TOVE85] Tovey, G.A. *Hill climbing with multiple local optima.* SIAM J. Alg. Disc. Math. Vol. 6, No. 3 (1985), 384-395.

[UEIN72] Ueing, U. *A combinatorial method to compute the global solution of certain non-convex optimization problems.* Numerical Methods for Non- Linear Optimization (eds. F.A.Lootsma Acad. Pr.1972) 223-230.

[VAIS74] Vaish, H. *Nonconvex programming with applications to production and location problems.* Ph. D. Thesis, Georgia Inst. of Technology (1974).

[VAIS76] Vaish, H. and Shetty, C.M. *The bilinear programming problem.* Naval Res. Logist. Quart. 23 (1976), 303-309.

[VAIS77] Vaish, H. and Shetty, C.M. *A cutting plane algorithm for the bilinear programming problem.* Naval Res. Logist. Quart. 24 (1977), 83-94.

[VASI84] Vasil'ev, N.S. *An active method of searching for the global minimum of a concave function.* USSR Comput. Maths. Math. Phys., Vol. 24, No. 1 (1984), 96-100.

[VASI83] Vasil'ev, N.S. *Search for a global minimum of a quasiconcave function.* USSR Comput. Maths. Math. Phys., Vol. 23, No. 2 (1983), 31-35.

[VASI85] Vasil'ev, N.S. *Minimum search in concave problems, using the sufficient condition for a global extremum.* USSR Comput. Maths. Math. Phys., Vol. 25, No. 1 (1985), 123-129.

[VEIN69] Veinott, A.F. *Minimum concave cost solution of Leontiev substitution models of multi-facility inventory systems.* Oper. Res. Vol. 14 (1969), 486-507.

[VEIN85] Veinott, A.F. *Existence and characterization of minima of concave functions on unbounded convex sets.* Mathm. Progr. Study 25 (1985), 88-92.

[VERI85] Verina, L.F. *Solution of some nonconvex problems by reduction to nonconvex parametric programming.* Vestsi Akad. Navuk BSSR Ser. Fiz. Mat. Navuk No.1 (1985), 13-18, 124.

[WATA84] Watanabe, H. *IC layout generation and compaction using mathematical programming.* Ph.D.thesis (1984), Computer Sc. Dept. Univ. of Rochester.

[WILD64] Wilde, D.J. *Optimum seeking methods.* Prentice Hall, Englewood Cliffs, N.J. (1964).

[WING85] Wingo, D.R. *Globally minimizing polynomials without evaluating derivatives.* Intern. J. Computer Math. Vol. 17 (1985), 287-294.

[ZALI78] Zaliznyak, N.F. and Ligun, A.A. *Optimal strategies for seeking the global maximum of a function.* USSR Comp. Math. Math. Phys. 18(2) (1978), 31-38.

[ZANG75] Zang, I. and Avriel, M. On functions whose local minima are global. JOTA Vol.16 (1975), 183-190.

[ZANG76] Zang, I. and Avriel, M. *A note on functions whose local minima are global.* JOTA Vol.18 (1976), 556-559.

[ZANG68] Zangwill, W.I. *Minimum concave cost flows in certain networks.* Manag. Sci. 14 (1968), 429-450.

[ZHIR85] Zhirov, V.S. *Search for the global extremum of a polynomial on a parallelepiped.* USSR Comput. Maths. Math. Phys., Vol. 25, No. 1 (1985), 105-116.

[ZILI82] Zilinskas, A. *Axiomatic approach to statistical models and their use in multimodal optimization theory.* Math. Progr. Vol. 22, No. 1(1982), 102-116.

[ZILV83] Zilverberg, N.D. *Global minimization for large scale linear constrained systems.* Ph.D.thesis, Computer Sc. Dept. Univ. of Minnesota (1983).

[ZWAR73] Zwart, P.B. *Nonlinear programming: Counterexamples to two global optimization algorithms.* Oper. Res. Vol.21, No.6 (1973), 1260-1266.

[ZWAR74] Zwart, P.B. *Global maximization of a convex function with linear inequality constraints.* Oper. Res. Vol.22, No.3 (1974), 602-609.

[ZWAR71] Zwart, P.B. *Computational aspects of the use of cutting planes in global optimization.* In "Proceeding of the 1971 annual conference of the ACM (1971), 457-465.